STARGAZING 2010

MONTH-BY-MONTH GUIDE TO THE NORTHERN NIGHT SKY

HEATHER COUPER & NIGEL HENBEST

www.philips-maps.co.uk

HEATHER COUPER and NIGEL HENBEST are internationally recognized writers and broadcasters on astronomy, space and science. They have written more than 30 books and over 1000 articles, and are the founders of an independent TV production company specializing in factual and scientific programming.

Heather is a past President of both the British Astronomical Association and the Society for Popular Astronomy. She is a Fellow of the Royal Astronomical Society, a Fellow of the Institute of Physics and a former Millennium Commissioner, for which she was awarded the CBE in 2007. Nigel has been Astronomy Consultant to *New Scientist* magazine, Editor of the *Journal of the British Astronomical Association* and Media Consultant to the Royal Greenwich Observatory.

Published in Great Britain in 2009 by Philip's,
a division of Octopus Publishing Group Limited
(www.octopusbooks.co.uk)
Endeavour House, 189 Shaftesbury Avenue,
London WC2H 8JG
An Hachette UK Company (www.hachette.co.uk)

Reprinted 2010

TEXT
Heather Couper and Nigel Henbest (pages 6–53)
Robin Scagell (pages 61–64)
Philip's (pages 1–5, 54–60)

ISBN 978-1-84907-045-4

Title page: The Pleiades (Michael Stecker/Galaxy)

ACKNOWLEDGEMENTS

All star maps by Wil Tirion/Philip's, with extra annotation by Philip's.
Artworks © Philip's.

All photographs courtesy Galaxy Picture Library:
Martin Lewis 8;
Robin Scagell 12, 20, 61 top, 62, 63, 64 top;
Dave Tyler 16;
Damian Peach 24;
Thierry Legault 28;
Nik Szymanek 32–33;
Yoji Hirose 36–37;
Ralf Vandebergh 41;
David Arditti 44, 64 bottom;
Pete Lawrence 48–49;
Michael Stecker 52–53;
Celestron International 61 bottom.

CONTENTS

The sight of diamond-bright stars sparkling against a sky of black velvet is one of life's most glorious experiences. No wonder stargazing is so popular. Learning your way around the night sky requires nothing more than patience, a reasonably clear sky and the 12 star charts included in this book.

Stargazing 2010 is a guide to the sky for every month of the year. Complete beginners will use it as an essential night-time companion, while seasoned amateur astronomers will find the updates invaluable.

THE MONTHLY CHARTS

Each pair of monthly charts shows the views of the heavens looking north and south. They are usable throughout most of Europe – between 40 and 60 degrees north. Only the brightest stars are shown (otherwise we would have had to put 3000 stars on each chart, instead of about 200). This means that we plot stars down to 3rd magnitude, with a few 4th-magnitude stars to complete distinctive patterns. We also show the ecliptic, which is the apparent path of the Sun in the sky.

USING THE STAR CHARTS

To use the charts, begin by locating the north Pole Star – Polaris – by using the stars of the Plough (see May). When you are looking at Polaris you are facing north, with west on your left and east on your right. (West and east are reversed on star charts because they show the view looking up into the sky instead of down towards the ground.) The left-hand chart then shows the view you have to the north. Most of the stars you see will be circumpolar, which means that they are visible all year. The other stars rise in the east and set in the west.

Now turn and face the opposite direction, south. This is the view that changes most during the course of the year. Leo, with its prominent 'sickle' formation, is high in the spring skies. Summer is dominated by the bright trio of Vega, Deneb and Altair. Autumn's familiar marker is the Square of Pegasus, while the winter sky is ruled over by the stars of Orion.

The charts show the sky as it appears in the late evening for each month: the exact times are noted in the caption with the chart. If you are observing in the early morning, you will find that the view is different. As a rule of thumb, if you are observing two hours later than the time suggested in the caption, then the following month's map will more accurately represent the stars on view. So, if you wish to observe at midnight in the middle of February, two hours later than the time suggested in the caption, then the stars will appear as they are on March's chart. When using a chart for the 'wrong' month, however, bear in mind that the planets and Moon will not be shown in their correct positions.

THE MOON, PLANETS AND SPECIAL EVENTS

In addition to the stars visible each month, the charts show the positions of any planets on view in the late evening. Other planets may also be visible that month, but they will not be on the chart if they have already set, or if they do not rise until early morning. Their positions are described in the text, so that you can find them if you are observing at other times.

We have also plotted the path of the Moon. Its position is marked at three-day intervals. The dates when it reaches First Quarter, Full Moon, Last Quarter and New Moon are given in the text. If there is a meteor shower in the month, we mark the position from which the meteors appear to emanate – the *radiant*. More information on observing the planets and other Solar System objects is given on pages 54–57.

Once you have identified the constellations and found the planets, you will want to know more about what's on view. Each month, we explain one object, such as a particularly interesting star or galaxy, in detail. We have also chosen a spectacular image for each month and described how it was captured. All of these pictures were taken by amateurs. We list details and dates of special events, such as meteor showers or eclipses, and give observing tips. Finally, each month we pick a topic related to what's on view, ranging from the Milky Way to double stars and space missions, and discuss it in more detail. Where possible, all relevant objects are highlighted on the maps.

FURTHER INFORMATION

The year's star charts form the heart of the book, providing material for many enjoyable observing sessions. For background information turn to pages 54–57, where diagrams help to explain, among other things, the movement of the planets and why we see eclipses.

Although there is plenty to see with the naked eye, many observers use binoculars or telescopes, and some choose to record their observations using cameras, CCDs or webcams. For a round-up of what's new in observing technology, go to pages 61–64, where equipment expert Robin Scagell shares his knowledge.

If you have already invested in binoculars or a telescope, then you can explore the deep sky – nebulae (starbirth sites), star clusters and galaxies. On pages 58–60 we list recommended deep-sky objects, constellation by constellation. Use the appropriate month's maps to see which constellations are on view, and then choose your targets. The table of 'limiting magnitude' (page 58) will help you to decide if a particular object is visible with your equipment.

Happy stargazing!

If ever there was a time to see A-list stars strutting their stuff, it's this month. The constellations of winter are so striking that there's no better time to start finding your way around the sky. And this year, they're joined by an interloper – the Red Planet **Mars**, which occupies centre stage in January. It joins the dazzling denizens of **Orion**, **Taurus**, **Gemini** and **Canis Major** to make up a scintillating celestial tableau.

▼ *The sky at 10 pm in mid-January, with Moon positions at three-day intervals either side of Full Moon. The star positions are correct for 11 pm at the*

JANUARY'S CONSTELLATION

You can't ignore **Gemini** in January. High in the south, the constellation is dominated by the stars **Castor** and **Pollux**, which are of similar brightness and represent the heads of a pair of twins – their stellar bodies run in parallel lines of stars towards the west. Legend has it that Castor and Pollux were twins, conceived on the same night by the princess Leda. On the night she married the King of Sparta, wicked old Zeus (Jupiter) invaded the marital suite, disguised as a swan. Pollux was the result of the liaison with Jupiter – and therefore immortal – while Castor was merely a human being. But the pair were so devoted to each other that Zeus decided to grant Castor honorary immortality, and placed both him and Pollux amongst the stars.

Castor, as it turns out, is an amazing star. It's actually not just one star, but a family of six. Even through a small telescope, you can see that Castor is a double star, comprising two stars circling each other. Both of these are double (although you need special equipment to detect this). Then there's also another outlying star, visible through a telescope, and this also turns out to be double.

PLANETS ON VIEW

Mars is the undoubted 'star' of the month, high in the south in the late evening skies and a brilliant beacon all night long.

eginning of January and 9 pm t the end of the month. The lanets move slightly relative o the stars during the month.

At the beginning of the year, the Red Planet shines at magnitude −0.8, but it rapidly brightens as Earth approaches, to reach a maximum brightness of magnitude −1.28 when Mars is at opposition on 29 January. The Red Planet starts the month near the Sickle of Leo, and moves rapidly westwards to end January just above the 'Beehive Cluster', Praesepe (M44), in Cancer.

Earlier in the evening, **Jupiter** – at magnitude −2.1 – dominates the horizon, low in the southwest on the borders of Capricornus and Aquarius and setting at around 7.30 pm. The much fainter **Neptune** (magnitude 8.0) starts the month in company with the giant planet – about two degrees to Jupiter's lower right – but the farthest planet slips more quickly into the twilight and is lost by the end of January.

Uranus lies rather higher in the west-southwest, on the borders of Aquarius and Pisces. On the limit of naked-eye visibility (magnitude 5.9), it sets at about 9.45 pm.

Ringworld **Saturn** (magnitude 0.8) lies in Virgo, and rises around 10.30 pm. A small telescope shows its rings pretty much edge on.

In the morning sky, look out for tiny **Mercury** low in the southeast during the second half of January: it's at greatest western elongation on 27 January. Mercury is rising at about 6.30 am, and during this fortnight it brightens from magnitude 0.5 to −0.1. **Venus** is too close to the Sun to be seen this month.

WEST

PISCES
21 Jan
TRIANGULUM
Mira
CETUS
Algol
PERSEUS
24 Jan
ARIES
Pleiades
Zeta
Capella
Zenith
AURIGA
Aldebaran
TAURUS
ERIDANUS
Rigel
LEPUS
27 Jan
Castor
Pollux
GEMINI
Betelgeuse
ORION
Sirius
CANIS MAJOR
COLUMBA
SOUTH
Procyon
CANIS MINOR
THE MILKY WAY
Adhara
PUPPIS
URSA MAJOR
Mars
30 Jan
CANCER
HYDRA
The Sickle
Regulus
3 Jan
LEO
VIRGO
Ecliptic
EAST
MS
SE

January's Object
Sirius

January's Picture
Mars

Radiant of
Quadrantids

Mars

Moon

MOON		
Date	Time	Phase
7	10.39 am	Last Quarter
15	7.11 am	New Moon
23	10.53 am	First Quarter
30	6.17 am	Full Moon

MOON

On the night of 2/3 January, the Moon passes below Praesepe and then Mars. On 4 January, it's near Regulus. In the pre-dawn hours of 7 January, the Last Quarter Moon lies between Spica (left) and Saturn (right). Early in the evenings of 17 and 18 January, you'll find the narrow crescent Moon near giant planet Jupiter. As the sky darkens on 25 January, the Pleiades (M45) lie just to the right of the Moon. On 29/30 January, the Moon is back with Mars and Praesepe and on 31 January it passes below Regulus.

SPECIAL EVENTS

On **3 January**, the Earth is at perihelion, its closest point to the Sun.

The maximum of the **Quadrantid** meteor shower also occurs on **3 January**. These shooting stars are tiny particles of dust shed by an old comet called 2003 EH1, burning up as they enter the Earth's atmosphere. Perspective makes them appear to emanate from one spot in the sky, the radiant (marked on the star chart). Unfortunately, this year bright moonlight will drown out the fainter meteors.

It won't be seen from the UK, but on **15 January**, there's an annular eclipse of the Sun – where the Moon moves right in front of Sun, but appears smaller, so a bright ring ('annulus') of the Sun's surface remains visible. The annular eclipse is visible from central Africa, the Indian Ocean, the south of India, Sri Lanka, Myanmar (Burma) and China. People in southeastern Europe, and most of Africa and Asia will be treated to a partial eclipse. More details at: http://eclipse.gsfc.nasa.gov/SEmono/ASE2010/ASE2010.html.

JANUARY'S OBJECT

This is the month of the brightest star in the sky – **Sirius**. It isn't a particularly luminous star: it just happens to lie nearby, at a distance of 8.6 light years. The 'Dog Star' is accompanied by a little companion, affectionately called 'The Pup'. This tiny star was discovered in 1862 by Alvan Clark when he was testing a telescope, but its existence had been predicted by Friedrich Bessel nearly 20 years before, when he observed that something was 'tugging' on Sirius. The Pup is a white dwarf: the dying nuclear reactor of an ancient star which has puffed off its atmosphere. White dwarfs are the size of a planet but have the mass of a star: because they're so collapsed, they have

⊙ Viewing tip

It may sound obvious, but if you want to stargaze at this most glorious time of year, dress up warmly! Lots of layers are better than a heavy coat – they trap air next to your skin – and heavy-soled boots stop the frost creeping up your legs. It may sound anorak-ish, but a woolly hat really does stop one-third of your body heat escaping through the top of your head. And – alas – no hip flask of whisky. Alcohol constricts the veins, and makes you feel even colder.

◄ *For this image Martin Lewis used a home-made Dobsonian telescope that he has adapted with a motor drive to follow the planet using a DMK webcam-based camera. He took separate video sequences through colour filters from his back garden in St Albans and combined the result on computer.*

considerable gravitational powers – hence Sirius' wobble. The Pup is visible through medium-powered telescopes.

JANUARY'S PICTURE

It has to be **Mars**! This image shows the Red Planet in all its glory. The dark patch at the centre is the Sinus Meridianus – Mars' equivalent of the Greenwich Meridian – while at left, you can see the planet's most prominent feature, Syrtis Major. Once they were thought to be vegetation, but we now know that these markings are simply dark rocks. At the bottom of the photo (inverted in this telescopic view) is the northern polar cap.

JANUARY'S TOPIC
Mars

On 29 January, Mars is at opposition, shining brighter than all the stars in the sky, bar Sirius (although Mars' elliptical orbit means it's actually closest to Earth – 99 million kilometres – two days earlier). Next to the Full Moon, and near Praesepe, the non-twinkling Red Planet will be unmistakeable. Use a small telescope to skim over its mottled surface and spot its icy polar caps.

The debate about ice, water and life on Mars has hotted up over the last few years. There's evidence from NASA's Viking landings in 1976 that primitive bacterial life was then still in existence. And the present flotilla of spaceprobes, which are crawling over the Red Planet's surface or orbiting it, are unanimously picking up evidence for present or past water all over Mars – the essential ingredient for life.

Next year sees the launch of a new mission to our neighbouring world – NASA's Curiosity rover (previously known as Mars Science Laboratory). It's a hugely ambitious project to build a rover – the size of a Mini Cooper – to investigate the geology, composition and potential for life of the Red Planet. For the first time, NASA is admitting that it is designing the project with a view towards the possibility of humans going to Mars.

The first signs of spring are on the way, as the winter star patterns start to drift towards the west, setting earlier. The constantly changing pageant of constellations in the sky is proof that we live on a cosmic merry-go-round, orbiting the Sun. Imagine it: you're in the fairground, circling on your horse, and looking out around you. At times you spot the ghost train, sometimes you see the roller-coaster and then you swing past the candy-floss stall. So it is with the sky – and the constellations – as we circle our local star. That's why we see different stars in different seasons.

FEBRUARY'S CONSTELLATION

One of the most ancient of the constellations, **Auriga** (the Charioteer), sparkles overhead on February nights. It is named after the Greek hero Erichthoneus, who invented the four-horse chariot in order to combat his lameness.

Auriga is dominated by **Capella**, the sixth brightest star in the sky. Its name means 'the little she-goat', but there's nothing little about Capella. The giant star is over 150 times more luminous than our Sun, and it also has a yellow companion.

Capella, the goat, marks the Charioteer's shoulder, and – to her right – is a tiny triangle of stars nicknamed 'the kids'. Two of the stars in the trio are variable stars – but not because of any intrinsic instability. They're 'eclipsing binaries', stars that change in brightness because a companion star passes in front of them. **Zeta Aurigae** is an orange star eclipsed every 972 days by a blue partner. **Epsilon Aurigae** is one of the weirdest star systems in the sky – see February's Topic.

And when Auriga's overhead, bring out your binoculars (better still, a small telescope) – for within the 'body' of the Charioteer are three very pretty star clusters, **M36**, **M37** and **M38**.

▼ The sky at 10 pm in mid-February, with Moon positions at three-day intervals either side of Full Moon. The star positions are correct for 11 pm at the

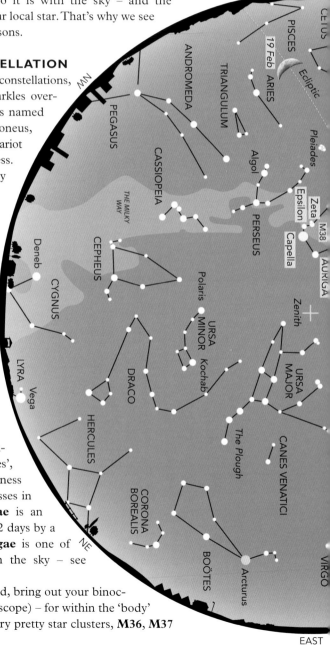

beginning of February and 9 pm the end of the month. The planets move slightly relative to the stars during the month.

PLANETS ON VIEW

Venus steps out of the dusk twilight this month: by the end of February, the Evening Star is setting at 6.55 pm, well over an hour after the Sun. Currently shining at magnitude −3.9, Venus is set to become brighter as it hangs in the evening sky right through to late summer.

By contrast, giant planet **Jupiter** (magnitude −2.0) is going downhill into the twilight glow: it's near Venus on 16 February (see Special Events).

Uranus (magnitude 5.9), on the borders of Aquarius and Pisces, sets at 8.45 pm at the start of February. Like Jupiter, it ends the month lost in the light of the setting Sun.

Mars is beginning to lose its lustre as the Earth pulls away from the Red Planet: during February, its magnitude drops from −1.3 to −0.6. It's visible all night long, in the constellation Cancer, near the cluster Praesepe (M44).

Around 8.30 pm, you'll find **Saturn** rising in Virgo. It's fainter than usual, at magnitude 0.7, as we're currently seeing the planet's bright rings almost edge on.

Both **Mercury** and **Neptune** are lost in the Sun's glare this month.

MOON

The Moon passes Saturn on the night of 2/3 February and Spica the following night. It's near the Pleiades (M45) on 21 February (see Special Events). On 25 February the Moon lies below Mars, and on 27 February it's near Regulus.

WEST

Zenith

PISCES · CETUS · PERSEUS · Pleiades · TAURUS · ERIDANUS · Epsilon · Zeta · M36 · 22 Feb · Aldebaran · LEPUS · M38 · M37 · Crab Nebula · Rigel · Capella · AURIGA · GEMINI · 25 Feb · Betelgeuse · ORION · Mirzam · CANIS MAJOR · Castor · Pollux · Procyon · Sirius · Adhara · CANIS MINOR · THE MILKY WAY · SOUTH · Mars · M67 · PUPPIS · URSA MAJOR · The Sickle · Praesepe · CANCER · Regulus · HYDRA · LEO · 28 Feb · VIRGO · Saturn · 2 Feb · Ecliptic · SE

EAST

Mars			
Saturn			
Moon			
February's Object Crab Nebula			
February's Picture Cancer			

MOON		
Date	**Time**	**Phase**
5	11.48 pm	Last Quarter
14	2.51 am	New Moon
22	0.42 am	First Quarter
28	4.38 pm	Full Moon

SPECIAL EVENTS

On **16 February**, the two brightest planets – Venus and Jupiter – pass less than a degree apart. This conjunction occurs very low in the west just before they set at 6 pm, and binoculars or a telescope will help to show the event.

As night falls on **21 February**, the First Quarter Moon brushes the fringes of the Pleiades – a lovely sight in binoculars or a small telescope.

FEBRUARY'S OBJECT

This month, we home in on **Taurus** (the Bull) – at a small region above his 'lower horn'. There, in 1054, Chinese astronomers witnessed the appearance of a 'new star', which outshone all the other stars in the sky. It was visible in daylight for 23 days and remained in the night sky for nearly two years. But this was no new star – it was an old star on the way out, which exploded because it was overweight.

Today, we see the remnants of the star as the **Crab Nebula** (M1) – so named by the 19th-century Irish astronomer Lord

▼ Robin Scagell captured this image of Cancer in February 1999, with a three-minute exposure on ISO 1600 film. He used a 50 mm lens at f/2.8, coupled with a diffusing filter, which helped to bring out the star colours. His observing site was the Chiltern Hills, to the west of London.

◉ Viewing tip

When you first go out to observe, you may be disappointed at how few stars you can see in the sky. But wait for around 20 minutes, and you'll be amazed at how your night vision improves. One reason for this 'dark adaption' is that the pupil of your eye gets larger to make the best of the darkness. More importantly, in dark conditions the retina of your eye builds up much bigger reserves of rhodopsin, the chemical that responds to light.

Rosse because it resembled a crab's pincers. Even today, the debris is still expanding from the wreckage and it now measures 15 light years across.

At the centre of the Crab Nebula is the core of the dead star, which has collapsed to become a pulsar. This tiny, but super-dense object – only the size of a city, but with the mass of the Sun – is spinning around furiously at 30 times a second and emitting beams of radiation like a lighthouse. You can just make out the Crab Nebula through a small telescope, but it is very faint.

FEBRUARY'S PICTURE

Although famous as a Zodiac constellation, **Cancer** (the Crab) contains no brilliant stars. But it does boast two important star clusters. At the top is **Praesepe**, bright enough to be just visible to the unaided eye; through binoculars, it looks like a swarm of bees – hence its common name, the 'Beehive Cluster'. While Praesepe is 600 million years old, Cancer's other cluster, **M67** (near the bottom left corner of the picture), is some six times older and has given astronomers valuable clues as to how stars age.

FEBRUARY'S TOPIC
Epsilon Aurigae

Star of the show this month is Epsilon Aurigae, a star 130,000 times brighter than the Sun, which is currently being eclipsed so that it appears only half its normal brightness. This two-year-long event happens every 27 years, and the current dimming will last until May 2011. But what's causing the eclipse? And why does Epsilon usually briefly brighten during mid-eclipse, due this August?

Whatever is doing the eclipsing must be huge – it would stretch beyond the orbit of Saturn if it were in our Solar System. The thinking at the moment is that it's a dark disc of dust surrounding one or two stars that orbit Epsilon itself, with their gravity acting as a 'vacuum cleaner' to keep the central region clear.

Whatever the nature of the beast may be, professional astronomers are desperate for observations from amateurs to help pin the creature down. Large telescopes today are too sensitive to deal with stars as bright as third magnitude, and amateurs, with modest telescopes – or even cameras – are in an ideal position to monitor the star's brightness. The American Association of Variable Star Observers is coordinating an Epsilon campaign: contact them at http://www.aavso.org/iya.

This month, the nights become shorter than the days as we hit the Vernal (Spring) Equinox – on 20 March, spring is 'official'. That's the date when the Sun climbs up over the equator to shed its rays over the northern hemisphere. Because of the Earth's inclination of 23.5° to its orbital path around the Sun, the north pole points away from our local star between September and March, causing the long nights of autumn and winter. Come the northern spring, Earth's axial tilt means that the Sun favours the north – and we can look forward to the long, warm days of summer.

You can spot four planets – **Venus, Mercury, Mars** and **Saturn** – in the evening sky, while the new season's constellations – like **Leo** and **Virgo** – are making their presence felt. And don't forget that the clocks go forwards on 28 March.

MARCH'S CONSTELLATION

Ursa Minor is a miniature version of **Ursa Major** – in fact, it's known as the 'Little Dipper' in America. It contains the most famous star in the sky: the Pole Star, or **Polaris** (see May's Object).

In legend, Ursa Minor was Arcas – the son of the Great Bear. Originally Ursa Major (mum) was Callisto – a beautiful nymph sworn to chastity. But the chief god Jupiter had other ideas. Their son was born in due course – and Juno, Jupiter's wife, was so furious that she had Callisto changed into a bear.

Years later, Arcas was out hunting with his father. In the forest, he saw a bear – and began to take aim. Horrified, Jupiter realized that she was the mother of Arcas and, using his godly powers, he changed Arcas into a bear. Then he swung mother and son into the heavens with such force, that their stumpy rumps turned into the elongated tails we see attached to the celestial bears today.

Ursa Minor is a useful model to show how to estimate stellar

▼ The sky at 10 pm in mid-March, with Moon positions at three-day intervals either side of Full Moon. The star positions are correct for 11 pm at the

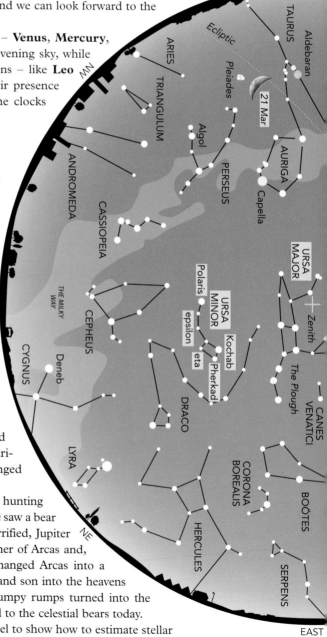

eginning of March and 10 pm *: the end of the month (after *ST begins).The planets move *ghtly relative to the stars during *e month.

brightness – and to gauge atmospheric transparency. **Polaris** is magnitude 2.1 (it varies slightly). Orange **Kochab** is the same brightness. **Pherkad** – a blue-white star – is at magnitude 3.1. **Epsilon Ursae Minoris** is a yellow giant with a brightness of magnitude 4.2. And the white star **eta** comes in last at magnitude 5.

PLANETS ON VIEW

If you haven't spotted **Venus** yet this year, you'll certainly be aware of its presence this month: by the end of March, the Evening Star is setting at 9.20 pm, almost two hours after the Sun and shining at magnitude –3.9.

Towards the end of March, Venus is joined by **Mercury**, speeding upwards in the dusk twilight. You can spot Mercury (magnitude –1.0), a few degrees to the lower right of the brilliant Evening Star, from 25 March onwards.

Mars is technically in Cancer all this month, but it's hanging around near the twin stars of Gemini, Cancer and Pollux. The Red Planet's brilliance plummets this month, from magnitude –0.6 to 0.2. It's visible most of the night, setting around 5.00 am.

Though overshadowed in brightness by Venus and Mars, this should really be the month of **Saturn**. The ringworld reaches opposition on 22 March, at magnitude 0.5, and shines all night long in Virgo. **Jupiter**, **Uranus** and **Neptune** are all too close to the Sun to be seen this month.

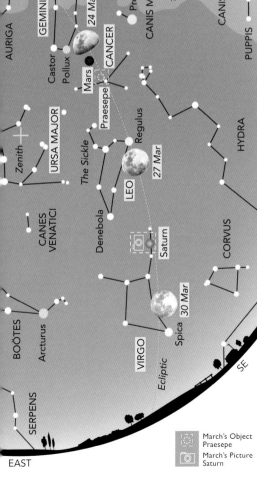

March's Object
Praesepe

March's Picture
Saturn

MOON		
Date	**Time**	**Phase**
7	3.42 pm	Last Quarter
15	9.01 pm	New Moon
23	11.00 am	First Quarter
30	3.25 am	Full Moon

Mars
Saturn
Moon

MOON

On the night of 1/2 March the Moon lies near Saturn and on 4 March it's just below Spica. In the early hours of 7 March, the Last Quarter Moon lies right next to Antares. At dusk on 17 March, look out for a narrow crescent Moon in the west, with Venus below. On 20 March, the Moon passes just below the Pleiades. The Moon is near Praesepe on 25 March, with Mars to the right. On 30 March, the Moon lies below Saturn.

SPECIAL EVENTS

The Vernal Equinox, on **20 March** at 5.32 pm, marks the beginning of Spring, as the Sun moves up to shine over the northern hemisphere.

28 March, 1.00 am: British Summer Time starts – don't forget to put your clocks on an hour (the mnemonic is 'Spring forwards, Fall back').

MARCH'S OBJECT

Between **Gemini** and **Leo** lies the faint zodiacal constellation of **Cancer**, the Crab. You'd be hard pressed to see it from a city, but try to concentrate your eyes on the central triangle of stars – then look inside. With the unaided eye, you can see a misty patch. This is **Praesepe** (M44) – a dense group of stars whose name literally means 'the manger', but is better known as the Beehive Cluster. If you train binoculars on it, you'll understand how it got its name – it really does look like a swarm of bees.

> ◉ **Viewing tip**
> This is the time of year to tie down your compass points – the directions of north, south, east and west as seen from your observing site. North is easy – just latch on to Polaris, the Pole Star. And before the clocks go forwards, the Sun is in the south at noon. But the useful extra in March is that we hit the Spring Equinox, when the Sun rises due east, and sets due west. So remember those positions relative to trees or houses around your horizon.

◄ Dave Tyler used his 355 mm C14 Schmidt-Cassegrain telescope from his home observatory in Flackwell Heath, Buckinghamshire, to capture this image of Saturn. The camera was a webcam-style Lumenera Skynyx.

Praesepe lies over 500 light years away, and contains some 300 stars, all of which were born together some 600 million years ago. Galileo, in 1610, was the first to recognize it as a star cluster. But the ancient Chinese astronomers obviously knew about it, naming the cluster Tseih She Ke – 'the Exhalation of Piled-up Corpses'!

MARCH'S PICTURE

Saturn is almost unreal when you see it through a small telescope. It's like a toy model, suspended in the blackness of space. Look at its belts of clouds, stretched out as a result of the planet's stupendous spin rate. This squashes the ringworld, making it look like a celestial tangerine.

Don't expect too much of the famed rings this year. Because of the relative positions of the Earth and Saturn, we are currently seeing them almost edge on. But Saturn's beautiful encircling halo will come back in a major way in the years to come, as our planet and the ringworld move on in their orbits.

MARCH'S TOPIC
Saturn

The slowly moving ringworld of Saturn is currently skulking in the peripheries of the sprawling constellation Virgo (the Virgin). It's famed for its huge engirdling rings, which would stretch nearly all the way from the Earth to the Moon.

And the rings are just the beginnings of Saturn's larger family. It has at least 60 moons, including Titan – where the international Cassini-Huygens mission has discovered lakes of liquid methane and ethane.

Saturn itself is second only to Jupiter in size. But it's so low in density that – were you to plop it in an ocean – it would float. Like Jupiter, Saturn has a ferocious spin rate – 10 hours and 32 minutes – and its winds roar at speeds of up to 1800 km/h.

Saturn's atmosphere is much blander than that of its larger cousin. But it's wracked with lightning bolts 1000 times more powerful than those on Earth.

Spring is here, with the skies dominated by the ancient constellations of **Leo** and **Virgo**. Leo does indeed look like a recumbent lion, but it's hard to envisage Y-shaped Virgo as a maiden holding an ear of corn!

APRIL'S CONSTELLATION

Like the mighty hunter Orion, **Leo** is one of the rare constellations that looks like its namesake – in this case, an enormous crouching lion. Leo is one of the oldest constellations and commemorates the giant Nemean lion that Hercules slaughtered as the first of his labours. According to legend, the lion's flesh couldn't be pierced by iron, stone or bronze – so Hercules wrestled with the lion and choked it to death.

The lion's heart is marked by the first-magnitude star **Regulus**, and its other end by **Denebola**, which means in Arabic 'the lion's tail'. A small telescope shows that **Algieba**, the star marking the lion's shoulder, is actually a beautiful close double star. Just underneath the main 'body' of Leo are several spiral galaxies – nearby cities of stars like our own Milky Way. They can't be seen with the unaided eye, but a sweep along the lion's tummy with a small telescope will reveal them.

PLANETS ON VIEW

Mercury is putting on its best evening appearance of the year, reaching greatest eastern elongation on 8 April. It starts the month at magnitude –0.8, but by the time it slips from sight around 17 April it has dropped to magnitude 2.0.

All this time, Mercury is keeping company with big sister Venus. While this may take the edge off Mercury's own performance, Venus acts a guide to finding the fainter planet. You'll have no problem spotting the brilliant Evening Star; Mercury is the brightish 'star' a few degrees to its right.

▼ *The sky at 11 pm in mid-April with Moon positions at three-day intervals either side of Full Moon. The star positions are correct for midnight at the beginning of April.*

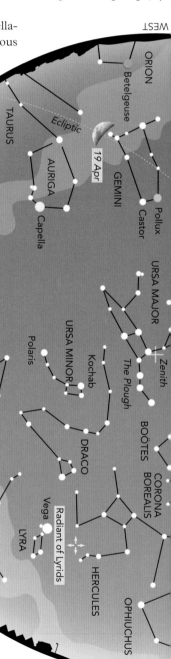

ind 10 pm at the end of the
month. The planets move slightly
elative to the stars during the
month.

Venus is the brightest object you'll see after sunset, bar the Moon, at magnitude −3.9. It climbs steadily up in the evening sky, in the company of Mercury for the first half of the month, passing below the Pleiades (M45) on 24 April.

Meanwhile, **Mars** is speeding eastwards through Cancer, passing Praesepe (M44) on 17 April. The Red Planet sets around 4 am and its magnitude drops from 0.2 to 0.8 during April.

Saturn, in Virgo, is shining all night long, at magnitude 0.7. Use a small telescope to view its rings – currently almost edge on – and its biggest moon, Titan (magnitude 8.3).

Jupiter, **Uranus** and **Neptune** are all lost in the Sun's glare in April.

MOON

In the early morning of 3 April, the Moon lies near Antares. The Moon passes near Mercury, Venus and the Pleiades on 15 and 16 April (see Special Events). On 21 and 22 April, the Moon lies near Mars and Praesepe and the following night it passes Regulus. On 25 April, the Moon passes below Saturn and for the following two nights it's near Spica. On the night of 30 April/1 May, the Moon is once again close to Antares.

SPECIAL EVENTS

There are two lovely sky sights mid-month: in the early evening of **15 April**, brilliant Venus lies to the upper left of the crescent Moon, with Mercury between them, and on the following evening – **16 April** – the Moon is immediately above Venus and below the Pleiades.

WEST

THE MILKY WAY

GEMINI
Procyon
CANIS MINOR
Castor
Pollux
CANCER
Mars
URSA MAJOR
The Sickle
Algieba
LEO
Regulus
22 Apr
HYDRA
Zenith
Denebola
Saturn
The Plough
CANES VENATICI
M87
Arcturus
Virgo Cluster
25 Apr
CORVUS
SOUTH
CORONA BOREALIS
BOÖTES
VIRGO
Spica
Ecliptic
28 Apr
HERCULES
SERPENS
LIBRA
OPHIUCHUS
EAST
SE
MS

April's Object
The Virgo Cluster

April's Picture
Leo

Mars

Saturn

Moon

MOON			
	Date	Time	Phase
	6	10.37 am	Last Quarter
	14	1.29 pm	New Moon
	21	7.20 pm	First Quarter
	28	1.18 pm	Full Moon

21/22 April: It's the maximum of the **Lyrid** meteor shower, which – by perspective – appears to emanate from the constellation of Lyra. The shower, which consists of particles from comet Thatcher, is active between 19 and 25 April. The best time to look for Lyrids is after 2 am, when the Moon has set.

APRIL'S OBJECTS

It's 'objects' this month – and big ones, too. If you have a small telescope, sweep the 'bowl' formed by Virgo's 'Y' shape, and you'll detect dozens of fuzzy blobs. These are just a handful of the thousands of galaxies making up the **Virgo Cluster**, our closest giant cluster of galaxies, lying at a distance of 55 million light years.

Galaxies are gregarious. Thanks to gravity, they like living in groups. Our Milky Way, and the neighbouring giant spiral the Andromeda Galaxy, live in a small cluster of more than 30 smallish galaxies called the Local Group.

▼ *Under the dark skies at Kelling Heath, Norfolk, Robin Scagell imaged the celestial lion. He made a nine-minute exposure using a Canon 10D DSLR at ISO 200, with the lens set at 27 mm. Kelling Heath is famed for its major star parties, where hundreds of amateur astronomers gather to observe and exchange information.*

But the Virgo Cluster is in a different league; it's like a vast galactic swarm of bees. What's more, its enormous gravity holds sway over the smaller groups around it – including our Local Group – to make a cluster of clusters of galaxies, the Virgo Supercluster.

The galaxies in the Virgo Cluster are also mega. Many of them are spirals like our Milky Way – including the famous 'Sombrero' (M104), which looks just like its namesake – but some are even more spectacular. The heavyweight galaxy of the cluster is **M87**, a giant elliptical galaxy emitting a jet of gas more than 4000 light years long that is travelling at one-tenth of the speed of light.

APRIL'S PICTURE

The constellation of **Leo** is dominated by his head and neck – the familiar '**sickle**', which looks like a back-to-front question mark. At the base of the sickle is the bright blue-white star Regulus. In this picture, you can compare its colour with the white star Denebola at far left and with the golden yellow of Algieba, the star halfway up the sickle. The star cluster at top left forms the constellation Coma Berenices.

APRIL'S TOPIC
Mercury

Mercury is putting in an unusually good appearance this month. The innermost planet, it zips around the Sun in only 88 days, and never strays far from it in the sky. So a glimpse of Mercury is quite rare – but now's your chance! (It's rumoured that Copernicus, the 'architect' of the Solar System, never observed the planet because of mists from the Vistula River in Poland.)

It looks like a bright, untwinkling star, low on the horizon. But don't expect any revelations, even through a good telescope. The planet is only a little larger than our Moon – and it bears an uncanny resemblance to it. Mercury is covered in craters and criss-crossed with wrinkled ridges like the skin of a dried-out apple – the result of the planet contracting after its hot birth.

NASA's Messenger spacecraft is on its way to Mercury, and is due to hit orbit on 18 March 2011. In its slowing-down manoeuvres, it has already taken new images of the pockmarked world. There are tantalizing hints that Mercury contains water in its parched rocks.

In 2013, Europe plans to launch an even more ambitious probe – BepiColumbo – which will map the planet thoroughly. It's named after an Italian scientist and technologist who worked out the 'gravity assist system' – how you could use one world's gravity to hurl a spacecraft on to the next.

Look south to find the red giant **Arcturus** – 'the Bearkeeper'. It's the brightest star in the constellation of **Boötes** (the Herdsman), who shepherds the two bears through the heavens. It's a sign that summer is on the way.

▼ The sky at 11 pm in mid-May, with Moon positions at three-day intervals either side of Full Moon. The star positions are correct for midnight at the beginning of May

MAY'S CONSTELLATION

Ursa Major – whose brightest stars are usually called '**The Plough**' – ties with Orion as being the most famous constellation. Orion's fame is clear to see: its stars are brilliant, making up a very powerful image of a giant dominating the autumn and winter sky. In contrast, those of the Plough are fainter, and most people today have probably never seen an old-fashioned horse-drawn plough, from which the constellation takes its name. In fact, some children call it 'the saucepan', while in America it's known as the Big Dipper.

But the Plough is the first constellation that most people get to know. There are two reasons. First, it's always on view in the northern hemisphere. And second, the two end stars of the 'bowl' of the Plough point directly towards the Pole Star, **Polaris** (see May's Object).

Though it seems so familiar, Ursa Major is unusual in a couple of ways. First, it contains a double star that you can actually split with the naked eye. **Mizar**, the star in the middle of the bear's tail (or the handle of the saucepan), has a fainter companion, **Alcor**.

And – unlike most constellations – most of the stars of the Plough lie at the same distance and were born together. Leaving aside the two end stars, **Dubhe** and **Alkaid**, the others are all moving in the same direction (along with brilliant Sirius, which is also a member of the group). Over thousands of years, the shape of the Plough will gradually change, as Dubhe and Alkaid go off on their own paths.

EAST

nd 10 pm at the end of the
nonth. The planets move slightly
elative to the stars during the
nonth.

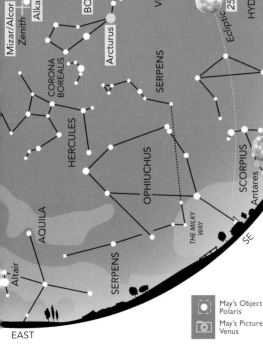

WEST

CANCER

Mars

19 May

The Sickle

Regulus

HYDRA

URSA MAJOR

LEO

Saturn

22 May

CORVUS

The Plough

CANES VENATICI

BOÖTES

Mizar/Alcor

Zenith

Alkaid

VIRGO

Spica

25 May

Ecliptic

HYDRA

CORONA BOREALIS

Arcturus

SERPENS

LIBRA

CENTAURUS

SOUTH

HERCULES

OPHIUCHUS

SCORPIUS

Antares

28 May

AQUILA

SERPENS

THE MILKY WAY

SE

Altair

EAST

PLANETS ON VIEW

The Evening Star **Venus** queens it over the western sky these May evenings, gradually brightening from magnitude –3.9 to –4.0. By the end of the month, Venus is setting just before midnight, so, rather unusually, it appears on our late evening charts. Venus starts the month near the Pleiades and races through Taurus to end the month in Gemini.

Mars is also heading eastwards, ahead of Venus, as it travels from Cancer to Leo, to finish the month near the celestial lion's brightest star, Regulus. During May, the Red Planet fades from magnitude 0.8 to 1.1 and by the end of the month it's setting at 1.45 am.

Ringworld **Saturn** is setting around 3.45 am. This magnitude 0.9 planet forms a gaudy trinket on one of the 'arms' of Virgo.

Towards the end of May, you may catch giant **Jupiter** rising in the east, in the morning twilight glow. It shines at magnitude –2.3 in Pisces.

Mercury is at greatest western elongation on 26 May, but it's lost in the dawn twilight – as are **Uranus** and **Neptune**.

MOON

The early morning of 1 May finds the Moon close to Antares. Before dawn on 9 and 10 May, the crescent Moon lies near Jupiter. On 15 and 16 May, there's a lovely composition in the evening sky of the Moon's narrow crescent with Venus. On 19 May, the Moon lies near Mars and the following night it's below Regulus. The Moon is near Saturn on 22 May and Spica on

Mars
Venus
Saturn

May's Object
Polaris
May's Picture
Venus

Moon

	MOON		
Date	Time	Phase	
6	5.15 am	Last Quarter	
14	2.04 am	New Moon	
21	0.42 am	First Quarter	
28	0.07 am	Full Moon	

23

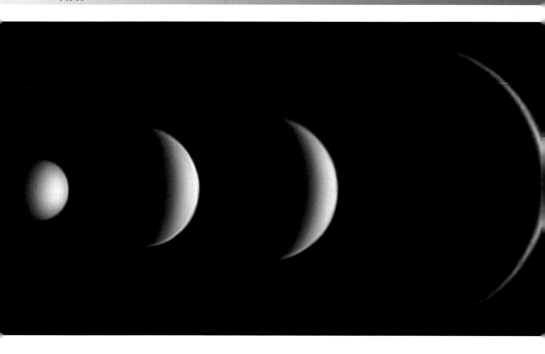

24 May. In the morning of 28 May, the Full Moon is once again up close and personal with Antares.

SPECIAL EVENTS

The maximum of the Eta Aquarid meteor shower falls on **5 May**, when tiny pieces of Halley's Comet burn up in Earth's atmosphere. Unfortunately, this year bright moonlight will drown out the fainter shooting stars.

MAY'S OBJECT

The Pole Star – **Polaris** – is a surprisingly shy animal, coming in at the modest magnitude of 2.1. You can find it by following the two end stars of **The Plough** (see chart) in **Ursa Major**, the Great Bear. Polaris lies at the end of the tail of the Lesser Bear (**Ursa Minor**), and it pulsates, making its brightness vary slightly over a period of four days. But its importance throughout recent history centres on the fact that Earth's north pole points to Polaris, so we spin 'underneath' it. It remains almost stationary in the sky, and acts as a fixed point for both astronomy and navigation. But over a 26,000-year period, the Earth's axis swings around like an old-fashioned spinning top – a phenomenon called precession – so our 'pole stars' change with time. Polaris will be nearest to the 'above pole' position in 2100, before the Earth wobbles off. Famous pole stars of the past include **Kochab** in Ursa Minor, which presided over the skies during the Trojan Wars of 1184 BC. In 14,000 years' time, brilliant **Vega**, in Lyra, will be our pole star.

> ⊙ *Viewing tip*
>
> Venus is a real treat this month. If you have a small telescope, though, don't wait too long to point it at the Evening Star. Once the sky gets dark, the cloud-wreathed planet appears so bright that it's difficult to make out anything on its disc. It's best to pick out Venus as soon as you can see it in the evening twilight, and you can then see the globe of Venus appearing against a pale blue sky.

◄ Damian Peach captured
these ghostly views of our
neighbour world Venus from
two different locations: Kent
and Tenerife. He compiled the
images, made in 2002, using
a webcam-type camera and
a 280 mm Schmidt-Cassegrain
telescope.

MAY'S PICTURE

This sequence of images captures the phases of our nearest world – **Venus**. As it orbits the Sun, we see it lit from a range of angles, just as we see our Moon. If you catch Venus through a small telescope this month, you'll see it between 'full' and 'half', like the gibbous Moon – as in the left-hand image of this sequence. Over the coming months, its diameter will appear to increase, even as its width shrinks to a thin crescent.

MAY'S TOPIC
Venus

As the Evening Star, Venus is unmissable this month. The planet of love is undeniably beautiful – it's pure white, and so brilliant that it can cast a shadow. But appearances are deceptive. Venus – almost the Earth's twin in size – is cloaked with dense clouds of carbon dioxide that reflect sunlight so efficiently that the planet literally dazzles.

Probe under those clouds, however, and you discover a world that has gone badly wrong. One commentator noted that – if you landed on Venus – you'd be simultaneously crushed, corroded, suffocated and roasted.

Crushed? That's the atmospheric pressure – 90 times that of the Earth. Corroded? Blame the sulphuric-acid rain in the upper atmosphere. Suffocated? Well – you can't breathe carbon dioxide. Roasted? Our nearest neighbouring world – two out from the Sun – is the hottest in the Solar System. With a surface boasting a temperature of 460°C, Venus is fiercer than an oven.

Almost certainly, active volcanoes are to blame, dumping vast amounts of carbon dioxide into the planet's atmosphere and leading to a runaway 'greenhouse effect'.

But Venus is a salutary example to *us* to reduce our carbon dioxide and methane footprints – which are of our own making.

This month should see the launch of Japan's Planet-C – a spaceprobe designed to investigate the planet's bizarre climate. And NASA has plans on the drawing-board for a lander that will confront the hell planet head-on: the Venus In-Situ Explorer.

On 21 June, the Sun hits the Summer Solstice – the date when our local star reaches its highest position over the northern hemisphere. This has been celebrated as a seasonal ritual for millennia, leading to the construction of massive stone monuments aligned on the rising Sun at midsummer. Undeniably, our ancestors had formidable astronomical knowledge.

▼ The sky at 11 pm in mid-June, with Moon positions at three-day intervals either side of Full Moon. The star positions are correct for midnight at the beginning of June.

JUNE'S CONSTELLATION

A tiny celestial gem, **Corona Borealis** rides high in the skies of early summer. In legend, it was the crown given as a wedding present by Bacchus to Ariadne. It really looks like a miniature tiara in the heavens, studded at its heart with the ultimate jewel – the blue-white star **Gemma** (magnitude 2.2). Gemma is a member of the Ursa Major association of stars (as is Sirius) and all of them move together in the same direction through space. Within the arc of the crown resides a remarkable variable star, **R Coronae Borealis**. It normally hovers around the limits of naked-eye visibility – sixth magnitude – but, unpredictably, it can drop to magnitude 14. That's because sooty clouds accumulate above the star's surface and obscure its light. The tiny crown also possesses another bizarre variable star, **T Coronae Borealis**, which behaves in the opposite way to its celestial compatriot. It usually skulks around at magnitude 11 (out of the range of binoculars), but then suddenly flares to magnitude 2. The 'Blaze Star' last erupted in 1946. It's what astronomers call a 'recurrent nova' – a white dwarf star undergoing outbursts after dragging material off its companion.

WEST

17 June

Ecliptic

Regulus

Mars

CANCER

The Sickle

Venus

14 June

Pollux

LEO

NW

Castor

GEMINI

URSA MAJOR

CANES VENATICI

The Plough

AURIGA

Zenith

Capella

HERCULES

NORTH

URSA MINOR

DRACO

Polaris

Vega

CASSIOPEIA

CEPHEUS

LYRA

PERSEUS

THE MILKY WAY

Algol

Deneb

CYGNUS

ANDROMEDA

NE

Square of Pegasus

PEGASUS

DELPHINUS

EAST

nd 10 pm at the end of the
onth. The planets move slightly
relative to the stars during the
onth.

PLANETS ON VIEW

After months of setting later and later, **Venus** is now slipping back down towards the Sun. Its early June bedtime of midnight is brought forward to 11.30 pm by the end of the month. The Evening Star blazes at magnitude –3.9, and during the month it travels from Gemini, through Cancer (passing Praesepe (M44) – see Special Events) to the borders of Leo.

Mars is ahead of Venus in the celestial track race, moving from the front of Leo through to the lion's hindquarters. It passes Regulus on 6 June (see Special Events). The Red Planet fades from magnitude 1.1 to 1.4 during June and by the end of the month it's setting just after midnight.

To the east (left) of Mars in Virgo, we find **Saturn** moving in the same direction – but the distant planet is a real plodder. The ringworld shines at magnitude 1.1, and sets at about 1.45 am.

Those with a telescope can now find **Neptune** (magnitude 7.9) in the morning sky on the borders of Capricornus and Aquarius. The most distant planet rises about 0.30 am.

And no one can miss **Jupiter**, the largest planet, shining at magnitude –2.4 in Pisces. It's rising at 2.30 am at the beginning of June, and as early as 0.30 am by the month's end. **Uranus** (magnitude 5.9) lies nearby; on 8 June, Jupiter passes just half a degree below Uranus, but the conjunction is low in the dawn twilight, and you'll need a telescope to spot this event.

Mercury is too close to the Sun to be visible this month.

WEST

Mars · 17 June · LEO · Saturn · VIRGO · 20 June · CORVUS · Spica · HYDRA · M5

URSA MAJOR · COMA BERENICES · CANES VENATICI · BOÖTES · Arcturus · LIBRA · 23 June · SCORPIUS · SOUTH

The Plough · Zenith · Gemma · CORONA BOREALIS · R · To · SERPENS · OPHIUCHUS · Ecliptic

DRACO · Vega · LYRA · HERCULES · SERPENS · Trifid Nebula · Lagoon Nebula · SAGITTARIUS · 26 June

CYGNUS · SAGITTA · Altair · AQUILA · SERPENS · CAPRICORNUS · THE MILKY WAY

PEGASUS · DELPHINUS · AQUARIUS · SE

EAST

June's Object
Vega

June's Picture
Lagoon and
Trifid Nebulae

Venus
Mars
Saturn
Moon

MOON		
Date	**Time**	**Phase**
4	11.13 pm	Last Quarter
12	12.14 pm	New Moon
19	5.29 am	First Quarter
26	12.30 pm	Full Moon

MOON

The morning of 6 June sees the crescent Moon hanging over Jupiter. On 14 and 15 June, the crescent forms a beautiful pair with Venus in the dusk sky. The Moon lies near to Mars and Regulus on 16 and 17 June. On 18 June, the Moon is below Saturn and on 20 June it's near Spica. In the mornings of 24 and 25 June, the almost-Full Moon lies near Antares.

SPECIAL EVENTS

Mars passes less than a degree from Regulus on **6 June**. They form a striking pair (enhanced by binoculars or a small telescope) with the steady orange glow of the planet contrasting with the twinkling blue-white star.

Around **10 June**, the Japanese spaceprobe Hayabusa should return to Earth after visiting minor planet Itokawa, carrying the first samples to be returned from an asteroid (though it's not known if the sample collection was in fact successful).

On **20 June**, Venus skims the upper edge of Praesepe (M44) – in almost exactly the position Mars was on 16 April.

▲ Thierry Legault used a Canon 10D DSLR for this stunning view from Angola. He added 22 separate five-minute exposures taken through a 106 mm Takahashi refractor.

21 June, 12.28 pm: Summer Solstice. The Sun reaches its most northerly point in the sky, so 21 June is Midsummer's Day, with the longest period of daylight. Correspondingly, we have the shortest nights.

There's a partial eclipse of the Moon on **26 June**, when the Full Moon is half obscured by the Earth's shadow. It's not visible from the UK, but will be on show to people on the Pacific Ocean and around the Pacific Rim.

JUNE'S OBJECT

One of our favourite stars – **Vega**, in **Lyra** – is rapidly ascending the heavens to occupy the zenith for the summer. The fifth brightest star in the sky, Vega has the honour to have been the first star photographed after the Sun. It is pure white, so pure

⊙ **Viewing tip**

This is the month for viewing the Sun – but be careful. NEVER use a telescope or binoculars to look at the Sun directly: it could blind you permanently. Fogged film is no safer, because it allows the Sun's infrared (heat) rays to get through. Eclipse goggles are safe (unless they're scratched). The best way to observe the Sun is to project its image through binoculars or a telescope on to a white piece of card.

that its colour is used as a benchmark to measure the colours of other stars – from red to blue-white – and so gauge their temperatures.

The star is a whirling dervish. It rotates in just 12.5 hours (as compared to roughly 30 days for our Sun), and – as a result – its equator bulges outwards, making Vega a tangerine-shaped star.

Vega was one of the first stars around which astronomers discovered a warm, dusty disc – like the one that formed our planetary system 4.6 billion years ago. There are strong suspicions that a planet with the mass of Jupiter might be lurking in there.

JUNE'S PICTURE

The Lagoon (bottom) and the Trifid nebulae are beautiful regions of starbirth in Sagittarius. The **Lagoon** (M8) lies more than 4000 light years away, and measures 100 light years across. The **Trifid** (M20) – so called because dark lanes of dust divide it into three – is also an active star factory. The black 'gaps' against the starry background are swathes of interstellar soot, poised to create future generations of stars.

JUNE'S TOPIC
Noctilucent clouds

Look north at twilight, and you may be lucky enough to see what has to be the most ghostly apparition in the night sky – noctilucent clouds. Their name is derived from the Latin for 'night shining' and these spooky clouds glow blue-white. Illuminated by the Sun from below the horizon, they're most commonly seen between latitudes 50° and 70° in the summer.

These are the highest clouds in the sky, occurring around 80 km up in the atmosphere. Their origin is controversial; they're certainly composed of ice, coated around tiny particles of dust – but what is the nature of the dust?

Tellingly, the first recorded observation of noctilucent clouds was made in 1885, two years after the eruption of Krakatoa. So could the particles be volcanic dust? Others believe that the dust could be micrometeorites, entering the atmosphere at high altitudes. Some scientists put them down to the industrial revolution, with its resultant increased pollution.

igh summer is here, and with it comes the brilliant trio of the Summer Triangle – the stars **Vega**, **Deneb** and **Altair**. Each is the brightest star in its own constellation: Vega in **Lyra**, Deneb in **Cygnus** and Altair in **Aquila**.

▼ *The sky at 11 pm in mid-July, with Moon positions at three-day intervals either side of Full Moon. The star positions are correct for midnight at the beginning of July*

JULY'S CONSTELLATION

Down in the deep south of the sky this month lies a baleful red star. This is **Antares** – 'the rival of Mars' – and in its ruddiness it even surpasses the Red Planet. To ancient astronomers, Antares marked the heart of **Scorpius**, the celestial scorpion.

According to Greek myth, this summer constellation is intimately linked with the winter star pattern Orion, who was killed by a mighty scorpion. The gods immortalized these two opponents as star patterns, set at opposite ends of the sky so Orion sets as Scorpius rises.

Scorpius is one of the few constellations that look like their namesakes. To the top right of Antares, a line of stars marks the scorpion's forelimbs. Originally, the stars we now call **Libra** (the Scales) were its claws. Below Antares, the scorpion's body stretches down into a fine curved tail (below the horizon on the chart), and deadly sting.

Scorpius is a treasure trove of astronomical goodies. Several lovely double stars include Antares; its faint companion looks greenish in contrast to Antares' strong red hue. Binoculars reveal the fuzzy patch of **M4** just to its west, a globular cluster made of tens of thousands of stars, some 7200 light years away.

The 'sting' contains two fine star clusters – **M6** and **M7** – so near to us that we can see them with the naked eye; a telescope reveals their stars clearly.

nd 10 pm at the end of the month. The planets move slightly relative to the stars during the month.

PLANETS ON VIEW

It's good and bad news for **Venus** fans. Good, because the planet is now brightening – from –4.0 to –4.2 this month – but bad because the Evening Star is slipping down into the twilight. By the end of July, it's setting only one and a half hours after the Sun. The planet is moving rapidly through Leo, passing a degree above Regulus on 19 July, in hot pursuit of Mars and Saturn. Through a small telescope, you'll see the illuminated part of Venus decrease as the solar illumination changes.

Mars, at magnitude 1.4, is now setting around 11.30 pm. The Red Planet spends the month heading eastwards through Leo and Virgo towards Saturn, catching up with the ringworld on the last day of August. Slowcoach **Saturn**, still in Virgo, shines at magnitude 1.1 and is now setting before midnight.

Giant planet **Jupiter**, in Pisces, is rising at 0.30 am at the beginning of July, and as early as 10.30 pm by the end of the month. Blazing at magnitude –2.6, Jupiter totally outshines its near neighbour in the sky, distant **Uranus** (magnitude 5.8), which lies a degree to the west (right) of Jupiter.

Neptune, in Aquarius, is even dimmer at magnitude 7.8, and rises at 10.30 pm.

Mercury is too close to the Sun to be seen in July.

MOON

On the morning of 4 July, the Last Quarter Moon hangs immediately above Jupiter, while the crescent

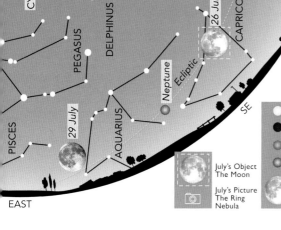

	Venus
	Mars
	Saturn
	Neptune
July's Object The Moon	
July's Picture The Ring Nebula	Moon

MOON		
Date	Time	Phase
4	3.35 pm	Last Quarter
11	8.40 pm	New Moon
18	11.10 am	First Quarter
26	2.36 am	Full Moon

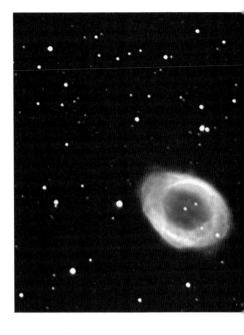

Moon forms a lovely pair with the Pleiades (M45) before dawn on 8 July. The waxing crescent Moon passes below Venus on the evening of 14 July, Mars on 15 July and Saturn on 16 July. The Moon is near Spica on 17 and 18 July. Antares is the star that keeps the Moon company on 21 July. On 30 July the waning Moon once again lies over Jupiter.

SPECIAL EVENTS

On **6 July**, the Earth is at aphelion, its furthest point from the Sun.

The European spaceprobe Rosetta will fly past the large asteroid Lutetia on **10 July**, sending back pictures and other data, on its way to rendezvous with Comet Churyumov-Gerasimenko in May 2014.

There's a total eclipse of the Sun on **11 July**, though it's not visible from Britain. The track of totality travels through the South Pacific from the Cook Islands past Easter Island to the very southern tip of South America. A partial eclipse is visible from all of the southern Pacific, Chile and Argentina. More details at: http://eclipse.gsfc.nasa.gov/SEmono/TSE2010/TSE2010.html.

JULY'S OBJECT

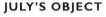

The **Moon** is our nearest celestial companion, lying a mere 384,400 km away. It took the Apollo astronauts only three days to reach it! And at 3,476 km across, it's so large compared to Earth that – from space – the system would look like a double planet.

But the Moon couldn't be more different from our verdant Earth. Bereft of an atmosphere, it has been exposed to bombardment by meteorites and asteroids throughout its life. Even with the unaided eye, you can see the evidence. The 'face' of the 'Man in the Moon' consists of huge, lava-filled craters created by asteroid hits 3.8 billion years ago.

Through binoculars or a telescope, the surface of the Moon looks amazing – as if you're flying over it. But don't observe our satellite when it's Full: the light is flat, and swamps its features. It's best to roam the Moon when it's a crescent, and see the sideways-on shadows highlighting its dramatic relief.

JULY'S PICTURE

The **Ring Nebula** in Lyra looks like a celestial smoke ring. It appears the same size as Jupiter in the sky, and William Herschel – who discovered Uranus – found many similar puzzling objects. He called them 'planetary nebulae'. But the Ring

⊙ **Viewing tip**

This is the month when you really need a good, unobstructed view to the southern horizon, to make out the summer constellations of Scorpius and Sagittarius. They never rise high in temperate latitudes, so make the best of a southerly view – especially over the sea – if you're away on holiday. A good southern horizon is also best for views of the planets, because they rise highest when they're in the south.

▲ *Under the balmy skies of La Palma in the Canary Islands, Nik Szymanek homed in on the Ring Nebula. He used a 250 mm Schmidt-Cassegrain telescope, coupled to an SBIG ST7 cooled CCD. The total exposure time was 45 minutes, through separate red, green and blue colour filters.*

Nebula is a dying star – a fate that will happen to our Sun in some 7 billion years' time. Its core has run out of nuclear fuel, and the unstable star has puffed off its outer layers into space. Eventually, these layers will disperse, leaving the core (centre) exposed as a cooling white dwarf star – which will later become a black, celestial cinder.

Lying between the two lowest stars of the tiny constellation, the Ring Nebula is very faint – magnitude 9. It's best to have a telescope with a mirror of around 200 mm to observe the planetary nebula well.

JULY'S TOPIC
Ophiuchus

If you're Moonwatching on the night of 22 July, you'll find it's hanging out in the constellation of Ophiuchus. But hang on – the Moon and planets usually sit around in the well-known signs of the zodiac, like Leo, Gemini or Taurus, surely? Not entirely true. When our familiar constellation patterns were drawn up in Mesopotamia and Greece more than 2000 years ago, astronomers noticed that the Sun, Moon and planets kept to a distinct band in the sky. They divided this special band into the constellations of the Zodiac, and assigned one star pattern for each month of the year. Misguided astrologers today still interpret the positions of the planets and the Sun in the Zodiac as omens for humankind – in particular, where the Sun is placed against the stars during your birth month. But think you're a Gemini? You're likely to be a Taurus. The reason for this is that everything has slipped back since the Greeks. Because of the gravity of the Moon, the Earth's axis wobbles with a period of 26,000 years – a phenomenon called 'precession', which changes the positions of the Sun and planets relative to the background stars. In addition, when you look in detail at a sky map, you'll find that this celestial highway crosses part of a 13th constellation – Ophiuchus. Alas: if you believe that you're a Sagittarian, you're probably really an Ophiuchian!

We have the Glorious Twelfth for astronomers this month: 12 to 13 August is the maximum of the **Perseid** meteor shower. This is the year's most reliable display of shooting stars and it also conveniently takes place during the summer, when it's not too uncomfortable to stay up late under the stars!

While you're at it, you can also take in a fine display of summer stars and – this year – a great gathering of planets in the west.

AUGUST'S CONSTELLATION

It may be small, but it's perfectly formed. **Delphinus**, the celestial dolphin, is outlined by four stars making a lopsided rectangle, with an extra star forming his tail. To find this constellation, first locate the Summer Triangle of the bright stars **Vega**, **Deneb** and **Altair**, then look to the upper left of Altair.

This constellation immortalizes humanity's long relationship with the most intelligent marine life on our planet. According to one myth, the dolphin acted as go-between when the sea god Poseidon (Neptune) was courting his wife, the sea nymph Amphitrite. In another story, the dolphin rescued the musician Arion when he was thrown overboard by sailors intent on stealing his wealth.

The two stars to the right of the rectangle are called **Sualocin** and **Rotanev**. These strange-looking names represent a bit of self-promotion by a 19th-century Italian astronomer, Niccolo Cacciatore. In Latin, his name becomes Nicolaus Venator: try spelling this backwards!

The top left star of Delphinus, **gamma Delphini**, is a lovely double star as you can see if you observe it with a reasonable telescope.

PLANETS ON VIEW

Early August sees a tremendous grouping of planets in the evening sky, though they will be low down in the twilight.

▼ *The sky at 11 pm in mid-August, with Moon positions at three-day intervals either side of Full Moon. The star positions are correct for midnight at the*

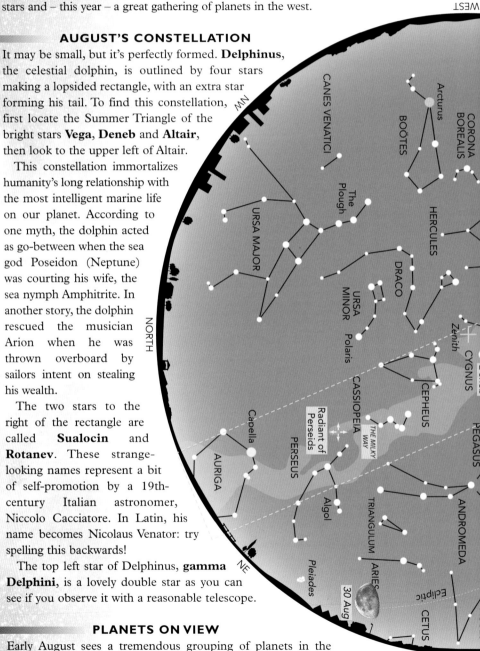

WEST

NORTH

NW

NE

EAST

CANES VENATICI
BOÖTES
Arcturus
CORONA BOREALIS
HERCULES
The Plough
URSA MAJOR
DRACO
URSA MINOR
Polaris
Zenith
CYGNUS
CEPHEUS
PEGASUS
Capella
Radiant of Perseids
CASSIOPEIA
THE MILKY WAY
AURIGA
PERSEUS
Algol
TRIANGULUM
ANDROMEDA
Pleiades
ARIES
30 Aug
Ecliptic
CETUS

beginning of August and 10 pm at the end of the month. The planets move slightly relative to the stars during the month.

WEST

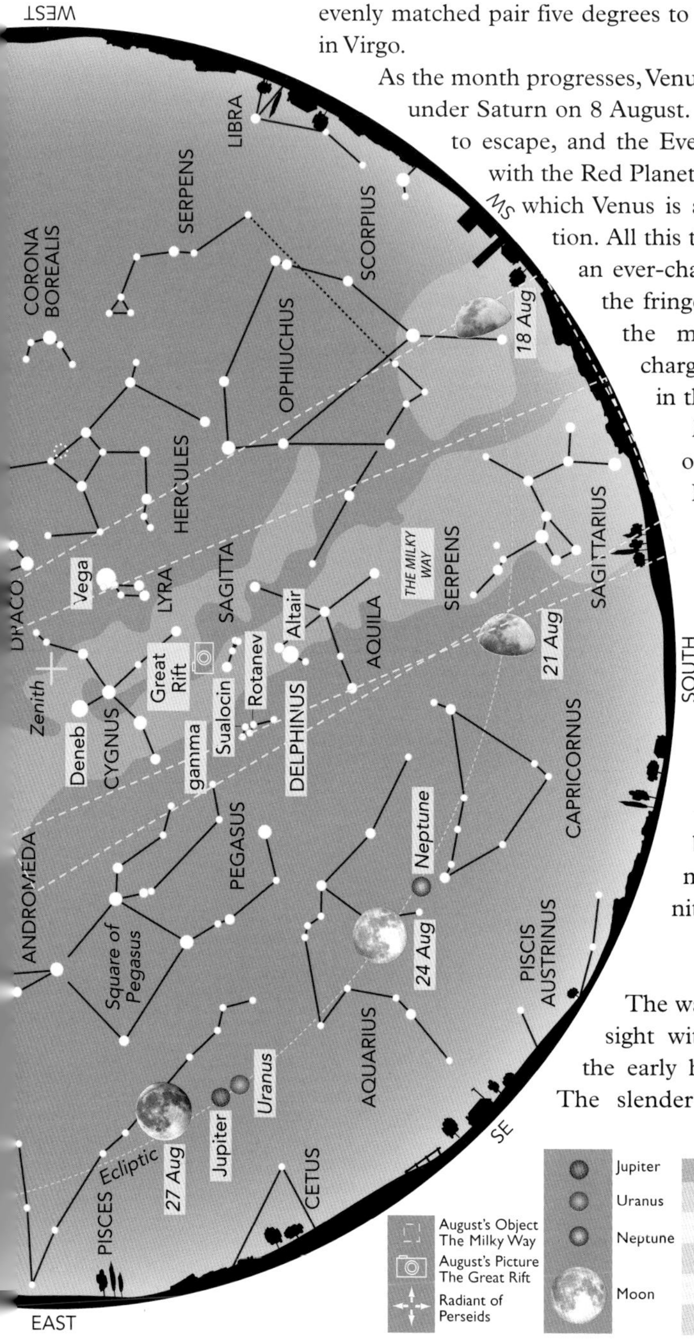

Technically, **Mercury** is part of this show as it's at greatest eastern elongation on 7 August, but it will be too low to be seen from Britain.

Venus is the most brilliant participant, at magnitude −1.3 and brightening. The month starts with Venus low in the west, and **Mars** (magnitude 1.5) and **Saturn** (1.1) forming an evenly matched pair five degrees to the upper left. All three lie in Virgo.

As the month progresses, Venus speeds eastwards, to pass under Saturn on 8 August. Meanwhile, Mars is trying to escape, and the Evening Star only catches up with the Red Planet on 20 August – the day on which Venus is at greatest eastern elongation. All this time, the three worlds form an ever-changing celestial triangle on the fringes of Virgo. For the rest of the month, Venus and Mars charge eastwards together – she in the lead – towards Spica.

Meanwhile, across the other side of the sky in Pisces, **Jupiter** is sedately shining at magnitude −2.8 and rising around 9.30 pm. **Uranus** (magnitude 5.8) lies three degrees to its right.

The most distant planet, **Neptune**, lies on the borders of Capricornus and Aquarius; it's at opposition on 20 August, but even so can only muster a brightness of magnitude 7.8.

MOON

The waning Moon forms a lovely sight with the Pleiades (M45) in the early hours of 4 and 5 August. The slender crescent Moon is near

August's Object
The Milky Way

August's Picture
The Great Rift

Radiant of
Perseids

Jupiter
Uranus
Neptune
Moon

MOON		
Date	Time	Phase
3	5.58 am	Last Quarter
10	4.08 am	New Moon
16	7.14 pm	First Quarter
24	6.04 pm	Full Moon

EAST

Venus, Mars and Saturn on 12 and 13 August (see Special Events). On 14 August, the Moon lies close to Spica. The Moon moves through Scorpius on 17 August, grazing past Alniyat (magnitude 2.9), the star to the right of Antares. On 26 and 27 August, the Moon lies near Jupiter. And on 31 August, the Moon is once again close to the Pleiades.

SPECIAL EVENTS

On **12 August**, the Moon lies near the great grouping of planets in Virgo – Venus, Mars and Saturn – adding to an already thrilling sight.

The maximum of the annual **Perseid** meteor shower falls on **12/13 August**. You'll see Perseid meteors for several nights around the time of maximum. This is an excellent year for observing the Perseids, as the Moon sets at 9 pm and the skies will be dark when the meteors reach their peak in the early hours.

AUGUST'S OBJECT

It's a stunning month for sweeping down the **Milky Way**, especially through binoculars. The stars look packed together, and you'll pick out star clusters and nebulae as you travel its length. These are all the more distant denizens of our local Galaxy, flattened into a plane because we live within its disc. It's akin to seeing the overlapping streetlights of a distant city on Earth.

But you'll notice something else – there is a black gash between the stars. William Herschel, the first astronomer to map the Galaxy, thought that this was a hole in space. But now we know that the **Great Rift** is a dark swathe of

sooty dust crossing our Galaxy's disc: material poised to collapse under gravity, heat up and – mixed with interstellar gas – create new generations of stars and planets. Proof that there is life in our old Galaxy yet!

AUGUST'S PICTURE

The **Great Rift** blocks the light of distant stars in this beautiful image of the summer **Milky Way**. At the top of the picture, you can pick out the three stars of the Summer Triangle: from left to right Deneb, Vega and (bottom) Altair. The rift wends its way through Cygnus and Aquila and ends near the horizon in the dense star clouds of Scutum and Sagittarius.

AUGUST'S TOPIC
Perseid meteor shower

So many people report to us that they see loads of shooting stars on their summer holidays – and are amazed when we observe, 'So you go on holiday in August.' But there's no mystery here. Between 8 and 12 August, Earth's orbit intersects a stream of debris from Comet Swift-Tuttle, which smashes into our atmosphere at speeds of 210,000 km/h and burns up. It forms the Perseids – the most reliable meteor shower of the year. Because of perspective, the meteors all appear to diverge from the same part of the sky – the *radiant* – which lies in the constellation Perseus. But there's no danger of being struck by a meteor. These tiny particles of dust (less than 1 cm across) burn up around 60 km above the Earth's surface. Look upon the shower as a celestial fireworks display – with the added bonus that August is much warmer than bonfire night!

◀ *From Japan's Mt Fuji – in April 1985 – Yoji Hirose captured the splendour of the Milky Way and the Great Rift. He used a Minolta SRT Super camera with a 16 mm lens at f/2.8 on Fujichrome 400D film. The exposure was ten minutes.*

◉ Viewing tip

Have a Perseids party! You don't need any optical equipment – in fact, telescopes and binoculars will restrict your view of the meteor shower. The ideal viewing equipment is your unaided eye, plus a sleeping bag and a lounger on the lawn. If you want to make measurements, a stopwatch and clock are good for timings, while a piece of string will help to measure the length of the meteor trail.

Autumn is here – with its unsettled weather – and we have the star patterns to match. **Aquarius** (the Water-carrier) is part of a group of aquatic star patterns which include **Cetus** (the Sea Monster), **Capricornus** (the Sea Goat), **Pisces** (the Fishes), **Piscis Austrinus** (the Southern Fish) and **Delphinus** (the Dolphin). There's speculation that the ancient Babylonians associated this region with water because the Sun passed through this zone of the heavens during the rainy season, from February to March.

SEPTEMBER'S CONSTELLATION

Aries – between Andromeda and Cetus – is not one of those constellations that grabs you. It has two moderately bright stars (**Hamal** and **Sheratan**), which – with fainter **Mesarthim** – make up the head of the celestial ram.

However, it's an ancient constellation. About 2000 years ago, the Sun – on its annual migration from the southern to the northern hemisphere – crossed the celestial equator in Aries. It was a celebration that spring was on the way, as the Sun climbed higher in northern skies.

In Greek mythology, this beast had an unfortunate ending. He was the 'Golden Ram' who rescued the hero Phrixos – only to be sacrificed to the gods by his saviour. Phrixos hung his skin in the temple, where it was coveted as the 'golden fleece'.

Of the three stars marking Aries' head, Mesarthim – the faintest – is the most interesting. It's a double star, consisting of two equally bright white stars easily visible through a small telescope.

PLANETS ON VIEW

The evening planets that have been with us for months are now sinking fast into the evening twilight glow: **Mars** and **Saturn** are now hard to spot without optical aid, and even Venus will have gone by the middle of September.

▼ The sky at 11 pm in mid-September, with Moon positions at three-day intervals either side of Full Moon. The star positions are correct for midnight at the

WEST

OPHIUCHUS
SERPENS
CORONA BOREALIS
Arcturus
CANES VENATICI
BOÖTES
HERCULES
The Plough
DRACO
Vega
LYRA
CYGNUS
Deneb
Zenith
CEPHEUS
URSA MAJOR
URSA MINOR
Polaris
CASSIOPEIA
ANDROMEDA
NORTH
AURIGA
Capella
THE MILKY WAY
PERSEUS
Algol
TRIANGULUM
ARIES
Pleiades
Ecliptic
26 Sept
29 Sept
Aldebaran
TAURUS
NE
EAST

eginning of September and
0 pm at the end of the month.
he planets move slightly relative
o the stars during the month.

During the first week of the month, you'll still spot **Venus** because of its brilliance (magnitude –4.4, rising during September to –4.6) even though it's very low in the west. Through a telescope, you'll see Venus rapidly grow in diameter this month, as its width shrinks to a crescent.

But there's no difficulty in finding **Jupiter**, which is now lording it over the night sky and is at opposition on 21 September. The giant planet is above the horizon all night long, shining at magnitude –2.9, in Pisces.

Uranus lies almost directly behind Jupiter, as seen from Earth, and comes to opposition a few hours later, on 22 September. The week from 15 to 22 September is an ideal time for spotting this faint world, on the limit of naked-eye vision (magnitude 5.7). Find Jupiter in binoculars or a small telescope (at low power), then move your view upwards by half a degree (one Moon's width). Uranus is the only reasonably bright 'star' you'll see – roughly the same brightness as Jupiter's main moons.

The most distant planet, **Neptune**, at magnitude 7.8, skulks on the borders of Capricornus and Aquarius, and sets around 4.15 am.

Mercury has the morning skies to itself, and is putting on its best dawn display of the year. During the second half of the month, look to the east about 5.30 am, and you'll see two 'stars'. The upper, fainter star is Regulus – the brighter object is Mercury, which reaches greatest western elongation on 19 September.

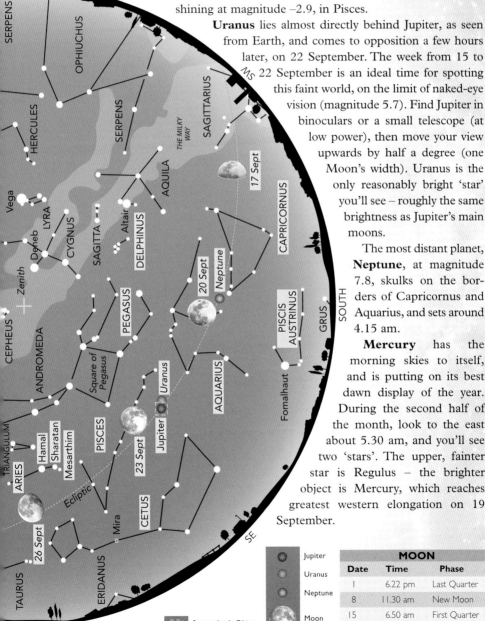

| | Jupiter |
| September's Object | Jupiter |

| | **MOON** | | |
Date	Time	Phase
1	6.22 pm	Last Quarter
8	11.30 am	New Moon
15	6.50 am	First Quarter
23	10.17 am	Full Moon

MOON

On 11 September, you'll see the thin crescent Moon to the left of Venus, low in the southwest after sunset. The Full Moon lies above Jupiter on 22 and 23 September. On the night of 27/28 September, the Moon passes below the Pleiades.

SPECIAL EVENTS

It's the Autumn Equinox at 4.09 am on **23 September**. The Sun is over the Equator as it heads southwards in the sky, and day and night are equal.

SEPTEMBER'S OBJECT

Jupiter is particularly bright this month. On 21 September, it's at 'opposition' – meaning that it's opposite the Sun in the sky, and at its closest to the Earth. 'Close' is a relative term, however – the planet is still over 590 million km away. But Jupiter is so vast – at 143,000 km in diameter, it could contain 1300 Earths – and as it's made almost entirely of gas, it's very efficient at reflecting sunlight.

Although Jupiter is so huge, it spins faster than any other planet in the Solar System. It rotates every 9 hours 55 minutes, and as a result its equator bulges outwards – through a small telescope, it looks a bit like a tangerine crossed with an old-fashioned humbug. The humbug stripes are cloud belts of ammonia and methane stretched out by the planet's dizzy spin.

Jupiter has a fearsome magnetic field that no astronaut would survive, huge lightning storms and an internal heat source, meaning that it radiates more energy than it receives from the Sun. Jupiter's core simmers at a temperature of 20,000°C.

Jupiter commands its own 'mini-solar system' – a family of more than 60 moons. The four biggest are visible in good binoculars, and even – to the really sharp-sighted – to the unaided eye. These are worlds in their own right – Ganymede is even bigger than the planet Mercury. But two vie for 'star' status. The surface of Io is erupting, with incredible geysers erupting plumes of sulphur dioxide 300 km into space. Brilliant white Europa probably contains oceans of liquid water beneath a solid ice coating, where alien fish may swim....

SEPTEMBER'S PICTURE

This incredible image of the **International Space Station** was captured from the firm ground of planet Earth. The ISS – permanently manned – orbits 350 km up, travelling at a speed of 28,000 km/h. The size of a football field, it's one of the brightest objects visible in the night sky, moving across the heavens at the speed of a high-flying plane.

New solar panels fitted in 2009 have made sightings of the ISS even more spectacular. These enormous arrays – 375 sq m

⊚ **Viewing tip**

Try to observe your favourite objects when they're well clear of the horizon. When you look low down, you're seeing through a large thickness of the atmosphere – which is always shifting and turbulent. It's like trying to observe the outside world from the bottom of a swimming pool! This turbulence makes the stars appear to twinkle. Low-down planets also twinkle – although to a lesser extent – because they subtend tiny discs, and aren't so affected.

► *Dutch amateur Ralf Vandebergh used a 250 mm Newtonian reflector with an Atik colour webcam for this shot of the ISS. He tracked the rapidly moving spacecraft by viewing through the telescope's finder, and selected the best frame from the resulting video stream.*

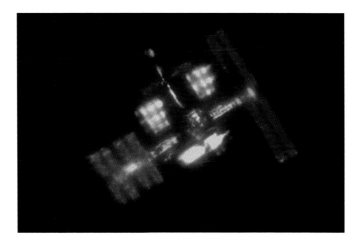

in area – may reflect sunlight and lead you to see a sudden dramatic flash in the sky. At times, the ISS can become 25 times more brilliant than Venus. More details at http://www.spaceweather.com.

SEPTEMBER'S TOPIC
Extrasolar planets

In the past decade, a revolution in astronomy has taken place. After years of fruitless searching, scientists have discovered around 300 planets circling nearby stars.

The first came in 1995, when Swiss astronomers Michel Mayor and Didier Queloz discovered that the faint star 51 Pegasi – just to the right of the great Square of Pegasus – was being pulled backwards and forwards every four days. It had to be the work of a planet, tugging on its parent star. Astonishingly, this planet is around the same size as the Solar System's giant, Jupiter, but it is far closer to its star than Mercury. Astronomers call such planets 'hot Jupiters'.

A team in California led by Geoff Marcy was already looking for planets and soon found more. In addition, astronomers have found several planets by watching for stars to dim in brightness as a planet passes in front. And one planet has even been discovered because its gravity is focusing the light from a very distant star.

All of these are big planets, because present-day telescopes on Earth can't locate a planet as small as our world. But all this has been set to change, since the launch in March 2009 of NASA's Kepler satellite. It is monitoring the brightness of stars to detect planets passing in front of their parent suns. And the hope is that – one day – it will find an Earth out there.

This month sees the appearance of the year's only predicted naked-eye comet, Hartley 2 – although an expected brilliant visitor is possible any time. British astronomer Malcolm Hartley discovered this steaming ball of celestial ices and dust in 1986, using the UK Schmidt telescope in Australia. **Comet Hartley 2** orbits the Sun in just six and a half years, coming in from Jupiter's orbit to a point just outside the Earth's orbit (see Special Events). But don't expect too much: sadly, bright moonlight will spoil the view. The comet's location is marked on the chart.

Jupiter is king of the night sky. Above it you'll find the barren Square of **Pegasus** riding high in the south, while below the giant planet there's a distinctly fishy lot of constellations swimming around the horizon.

OCTOBER'S CONSTELLATION

Piscis Austrinus (the Southern Fish) lies low in the southwest. It's hardly a compelling constellation. But it has fond childhood memories for both of us, because this is the only time of year that we got to see its brightest star, **Fomalhaut**: the southernmost first-magnitude star visible from Britain.

Its name – derived from the Arabic – means 'the mouth of the whale'. But one of its nicknames seems more appropriate – 'the lonely star of Autumn'.

But Fomalhaut isn't really lonely. It has a disc of debris in orbit about it, with the potential to form a planetary family. And in 2008, the Hubble Space Telescope captured an image of a world within the disc – the first to be seen beyond the Solar System.

PLANETS ON VIEW

Mighty **Jupiter** rules the planetary scene right through to its setting at 5 am, blazing at magnitude –2.9 in the dim constellation of Pisces. Use a small telescope, or even binoculars

▼ The sky at 11 pm in mid-October, with Moon positions at three-day intervals either side of Full Moon. The star positions are correct for midnight at the

WEST

OPHIUCHUS

AQUILA

CORONA BOREALIS

HERCULES

LYRA

THE MILKY WAY

CYGNUS

Vega

BOÖTES

DRACO

Deneb

CANES VENATICI

The Plough

URSA MINOR

CEPHEUS

Zenith CASSIOPEIA

Polaris

Algol

PERSEUS

NORTH

URSA MAJOR

Comet Hartley 2 15 Oct

Capella 20 Oct

AURIGA

Castor

GEMINI

25 Oct

Radiant of Orionids

26 Oct

Aldebaran

Betelgeuse

ecliptic

Pollux

29 Oct

30 Oct

ORION

NE

EAST

eginning of October and 9 pm
t the end of the month (after
te end of BST).The planets move
ightly relative to the stars during
te month.

supported steadily, to watch the perpetual dance of Jupiter's four largest moons.

You'll find **Uranus** about two degrees to the upper left of Jupiter, at the verge of naked-eye vision at magnitude 5.7. (Don't be fooled by the star 20 Piscium which is the same brightness and lies closer to Jupiter.)

Fainter **Neptune** is still hanging around between Capricornus and Aquarius. At magnitude 7.8, it's now setting around 2 am.

In the last week of October, **Saturn** creeps up into the morning sky: look for the ringed planet in the east around 5 am. It's shining at magnitude 0.9 in Virgo.

Mercury, **Venus** and **Mars** are too close to the Sun to be seen this month.

MOON

The waning crescent Moon lies near Regulus on the morning of 5 October and waxing crescent is near Antares after sunset on 11 October. On 19 and 20 October, you'll find the Moon sailing over Jupiter. The Moon lies near the Pleiades on 24 and 25 October. On 27 and 28 October, the Moon passes Comet Hartley 2 (see Special Events), spoiling our view of that rare visitor!

SPECIAL EVENTS

On **11 October**, NASA's EPOXI mission (originally called Deep Impact) is scheduled to fly past and take close-up pictures of Comet Hartley 2 (see next entry).

Comet Hartley 2 is closest to the Earth and at its brightest on

		October's Object
		Mira
		October's Picture
		M33
		Radiant of
		Orionids

Jupiter
Uranus
Neptune
Moon

MOON		
Date	**Time**	**Phase**
1	4.52 am	Last Quarter
7	7.44 pm	New Moon
14	10.27 pm	First Quarter
23	2.36 am	Full Moon
30	1.46 pm	Last Quarter

WEST

SERPENS
THE MILKY WAY
AQUILA
14 Oct
MS
CAPRICORNUS
Altair
SAGITTA
CYGNUS
Deneb
DELPHINUS
Enif
AQUARIUS
17 Oct
Neptune
PISCIS AUSTRINUS
CASSIOPEIA
Zenith
Andromeda Galaxy
PEGASUS
Square of Pegasus
Uranus
Jupiter
Fomalhaut
SOUTH
PERSEUS
Almach
ANDROMEDA
M33
20 Oct
PISCES
Ecliptic
CETUS
Algol
TRIANGULUM
ARIES
23 Oct
Mira
Pleiades
26 Oct
Aldebaran
TAURUS
ERIDANUS
SE
Betelgeuse
ORION
Rigel

EAST

20 October, as it skims past our planet – just outside Earth's orbit – on the way to its closest point to the Sun (28 October). We are viewing the comet almost end-on so don't expect to see much of a tail. Comet Hartley 2 should reach naked-eye brightness, around magnitude 4, although bright moonlight may mean you'll need binoculars to see the comet well.

Debris from Halley's Comet smashes into Earth's atmosphere on **21 October**, causing the annual **Orionid** meteor shower. This year moonlight will drown the fainter shooting stars.

At 2 am on **31 October**, we see the end of British Summer Time for this year. Clocks go backwards by an hour.

OCTOBER'S OBJECT

In another aquatic constellation close to Piscis Austrinus – **Cetus** (the Whale) – we find another extraordinary star. It was first observed in 1596 by the German astronomer David Fabricus. He thought it was an exploding star, because it flared up and then vanished. But – less than a year later – it was back. Fabricus had discovered the first known variable star, and called it **Mira** – Latin for 'the wonderful'.

Mira is a foretaste of what will befall our Sun, 7 billion years hence. It has swollen into a bloated red giant star so vast that – were it placed in our Solar System – it would engulf all the planets out to the asteroid belt.

It has no gravitational control over its girth. Mira swells and shrinks over a period of 332 days, rising to a bright magnitude 2 at maximum, and only magnitude 10 at minimum, when you need a medium-sized telescope to see it at all.

⊙ **Viewing tip**

Around 2.5 million light years away from us, the Andromeda Galaxy is often described as the furthest object easily visible to the unaided eye. But it's not *that* easy to see – especially if you are suffering some light pollution. The trick is to memorize the star patterns in Andromeda and look slightly to the *side* of where you expect the galaxy to be. This technique – called 'averted vision' – causes the image to fall on parts of the retina that are more light sensitive than the central region, which is designed to see fine detail.

◀ *Unbelievably, this photo of the faint and elusive galaxy M33 was taken from Edgware, in the suburbs of London. David Arditti used a 100 mm refractor. He compiled the image from a total of eight hours of separate exposures through colour filters and a light-pollution filter using a monochrome CCD camera.*

Mira is charging through space at a reckless rate, and – as a result – the unstable star is trailing a tail of gas 13 light years long. But this isn't all bad news. The elements in Mira's tail have the power to kickstart new stars and planets into being born.

Look out for Mira between 21–31 October this year, when it's due to reach maximum brightness.

OCTOBER'S PICTURE

The galaxy **M33** in **Triangulum** is the third most luminous galaxy in our Local Group – our local community of the cosmos, which comprises around 30–40 galaxies. But you'd be hard pressed to see M33. Although it hovers around the limit of naked-eye brightness (magnitude 6), it has a very low surface luminosity. We've heard that it can be seen with the unaided eye from remote desert locations, but don't try it from London!

M33 lies about 3 million light years away (a little farther off than the **Andromeda Galaxy**) and now is the time to try to spot it – preferably with a medium-sized telescope. It's about half the size of our Milky Way, but its spiral arms are much more ragged. The galaxy is home to one of the biggest star-forming regions known – NGC 604, which is 1500 light years across.

OCTOBER'S TOPIC
Light pollution

We once received a letter from a lady in Kent, saying, 'Before the War, we could see so many stars. But they're not there any more. Have they faded?' No – light pollution is the culprit. Our skies are crammed with particles of dust, coming from sources like car exhausts and factory emissions. Couple this with badly designed street lighting – and, hey presto, the stars disappear. It's not just an aesthetic issue. It's calculated that Britain throws away £52 million and two power stations' worth of energy a year in badly designed lighting. It is robbing us of our vision of the skies – and contributing to global warming. It isn't even making our lives any safer from crime, for research shows that more crimes are committed in well-lit areas. What's to be done? The Clean Neighbourhoods Act, which addresses – among other nuisances – noise and light pollution, should make a difference. In addition, lighting engineers are actively working on new designs for streetlights that point the light down, and not up. We hope this will bring us our vision of the dark night sky again. Otherwise, the only recourse will be to sit under the artificial skies of a planetarium.

When we see the **Pleiades** (M45), we know that winter is upon us. 'A swarm of fireflies tangled in a silver braid' was the evocative description of the glittering star cluster by Alfred, Lord Tennyson, in his 1842 poem 'Locksley Hall'. Throughout history, and all over the world, people have been intrigued by this lovely sight. Amazingly, from Greece to Australia, ancient myths independently describe the stars as a group of young girls being chased by an aggressive male – often **Aldebaran** or **Orion**, thus giving rise to the cluster's popular name, the Seven Sisters. Polynesian navigators used the Pleiades to mark the start of their year and farmers in the Andes rely on the visibility of the Pleiades as a guide to planting their potatoes: the brightness or faintness of the Seven Sisters depends on El Niño, which affects the forthcoming weather.

NOVEMBER'S CONSTELLATION

Taurus is very much a second cousin to brilliant **Orion**, but a fascinating constellation nonetheless. It is dominated by **Aldebaran**, the baleful blood-red eye of the celestial bull. Around 68 light years away and shining at a magnitude of 0.85, Aldebaran is a red giant star, but not one as extreme as neighbouring **Betelgeuse**. It is around three times heavier than the Sun. The 'head' of the bull is formed by the **Hyades** star cluster. The other famous star cluster in Taurus is the far more glamorous **Pleiades**, whose stars – although further away than the Hyades – are younger and brighter.

Taurus has two 'horns' – the star **El Nath** (Arabic for 'the butting one') to the north and **zeta Tauri** (which has an unpronounceable Babylonian name meaning 'star in the bull towards the south'). Above this star is a stellar wreck – literally. In 1054, Chinese astronomers witnessed a brilliant 'new star' appear in

▼ *The sky at 10 pm in mid-November, with Moon positions at three-day intervals either side of Full Moon. The star positions are correct for 11 pm at the*

beginning of November and
9 pm at the end of the month.
The planets move slightly relative
to the stars during the month.

this spot, which was visible in daytime for weeks. What the Chinese actually saw was an exploding star – a supernova – in its death throes. And today, we see its still-expanding remains as the **Crab Nebula**. It's visible through a medium-sized telescope.

PLANETS ON VIEW

You can't miss mighty **Jupiter**, dominating the dull skies of autumn on the borders of Aquarius and Pisces at magnitude –2.7. It's high in the southwest in the evening, and only sinks beneath the horizon at 2 am.

The seventh planet, **Uranus**, is just visible to the naked eye, at magnitude 5.8. It's about two degrees to the upper left of Jupiter, forming the top vertex of a triangle with two stars of similar brightness, 20 and 24 Piscium.

Neptune is skulking at magnitude 7.9 in Capricornus. It's now setting at around 11 pm.

Saturn rises at around 3 am. At magnitude 0.9, it's brightening up a dim region of the giant constellation Virgo. Through a small telescope, you'll notice its rings are considerably wider than they were earlier in the year.

And **Venus** is roaring into view as the Morning Star. Invisible at the beginning of November, by the end of the month it's rising three and a half hours before the Sun. At the same time, it brightens to an amazing magnitude –4.7. A small telescope – or even good binoculars – will show Venus as a lovely crescent.

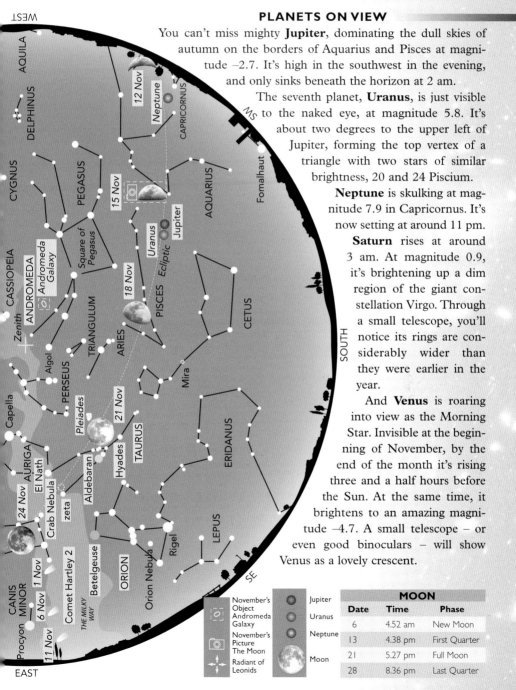

Legend		
November's Object Andromeda Galaxy	●	Jupiter
	●	Uranus
November's Picture The Moon	●	Neptune
Radiant of Leonids		Moon

MOON		
Date	Time	Phase
6	4.52 am	New Moon
13	4.38 pm	First Quarter
21	5.27 pm	Full Moon
28	8.36 pm	Last Quarter

Mercury and Mars are hidden in the Sun's glare this month.

MOON

On the morning of 4 November, the thin crescent Moon lies below Saturn. You'll find the Moon above Jupiter on 16 November. On 21 November, the Full Moon passes immediately below the Pleiades. The Last Quarter Moon is near Regulus on 27/28 November.

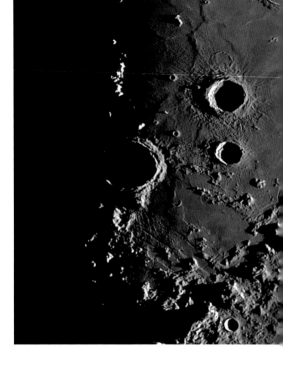

SPECIAL EVENTS

The **first week of November** is a good time for spotting **Comet Hartley 2**. Although the comet's now fading as it moves away from the Earth and Sun, the Moon has at last gone and we can see the comet in a dark sky. Observe in the wee hours of the morning when the comet has risen well above the horizon. Binoculars should give a grand view through to the end of the month.

17 November sees the maximum of the **Leonid** meteor shower. A few years ago, this annual shower yielded literally storms of shooting stars, but the rate has gone down as the parent comet Tempel-Tuttle, which sheds its dust to produce the meteors, has moved away from the vicinity of Earth. This year, bright moonlight will interfere.

NOVEMBER'S OBJECT

Take the advantage of autumn's newborn darkness to pick out one of our neighbouring galaxies – **M31** in **Andromeda**. M31 (the **Andromeda Galaxy**) is easily visible to the unaided eye from a dark location. It covers an area about four times the diameter of the Full Moon. Like our Milky Way, it is a beautiful spiral shape, but – alas – it's presented to us almost edge on.

The Andromeda Galaxy lies around 2.5 million light years away, and it's similar in size and shape to the Milky Way. It also hosts two bright companion galaxies – just like our Milky Way – as well as a flotilla of orbiting dwarf galaxies.

Unlike other galaxies, which are receding from us – as a result of the expansion of the Universe – the Milky Way and Andromeda are approaching each other. It's estimated that they will merge in 5 billion years' time. The result of the collision will be a giant elliptical galaxy – devoid of the gas that gives birth to stars – and dominated by ancient red stars.

⊙ **Viewing tip**

Now that the nights are drawing in earlier, and becoming darker, it's a good time to pick out faint, fuzzy objects like the Andromeda Galaxy and the Orion Nebula. But don't even think about it near the time of Full Moon – its light will drown them out. The best time to observe 'deep-sky objects' is when the Moon is near to New, or after Full Moon. Check the Moon phases timetable in the box.

NOVEMBER'S PICTURE

Our pockmarked **Moon** is home to thousands of craters. Although the Earth was also heavily bombarded by meteorites in the past, our outer planet has eroded the scars. In this image, the half-shaded crater on the left is Archimedes (83 km in diameter) in the Moon's Mare Imbrium. Its floor is very smooth, having been flooded with lava. To its right (top) is Aristillus (55 km in diameter), and Autolycus (39 km in diameter) lies below. All three are named after Greek astronomer-mathematicians. To the far right lie the peaks of the Montes Apenninus, marking part of the wall of Mare Imbrium. The range stretches 600 km, and some of the mountains are over 5000 m in height. NASA's Apollo 15 mission landed at the foot of this range.

▲ *Pete Lawrence, observing from Selsey in Sussex, got up close and personal with the Moon in this shot. He took it at the prime focus of a C14 380 mm Schmidt-Cassegrain telescope using a Lumenera Skynyx webcam-type camera.*

NOVEMBER'S TOPIC
Supernovae

With the star-wreck of the Crab Nebula sailing in the sky, it's a sobering reminder that some stars die young – and violently – with all the repercussions it will have on their planets and their lifeforms.

Supernovae – exploding stars – have been logged since ancient times. The Chinese called them 'guest stars'.

Stars more than eight times more massive than the Sun are doomed to an early death. They rip through the nuclear reactions that power their energy at a reckless rate. While our modest Sun converts hydrogen to helium in its core – making it shine – the biggies are far more ambitious. When the helium has run out, their gravity squeezes tighter, building successively new elements in the star's central nuclear reactor. All goes well – the star stays shining – until the core is made of iron.

Then it tries to fuse iron. It's a fatal mistake. Iron fusion takes *in* energy – and, as a result, the core collapses catastrophically. The star can't stand up to the shock. A burst of neutrinos blasts through its outer layers, blowing the star apart.

At its maximum brightness, a supernova can outshine a whole galaxy of 100,000 million stars.

The supernova hurls a cornucopia of elements into space – those from inside the dead star, and others created in the inferno of the explosion. But – in the end – a supernova is a phoenix, for out of its ashes, the seeds of life will arise.

This month sees the shortest day – and the longest night. On 21 December, we hit the Winter Solstice – the nadir of the year, which has long been commemorated in tablets of stone aligned to welcome the returning Sun. But the darkness brings with it a welcome fireworks display of meteors on 13 December, as the Geminids streak into our atmosphere.

▼ The sky at 10 pm in mid-December, with Moon positions at three-day intervals either side of Full Moon. The star positions are correct for 11 pm at the

DECEMBER'S CONSTELLATION

Spectacular **Orion** is one of the rare star groupings that looks like its namesake – a giant of a man with a sword below his belt, wielding a club above his head. Orion is fabled in mythology as the ultimate hunter.

The constellation contains one-tenth of the brightest stars in the sky: its seven main stars all lie in the 'top 70' of brilliant stars. Despite its distinctive shape, most of these stars are not closely associated with each other – they simply line up, one behind the other.

Closest is the star that forms the hunter's right shoulder, **Bellatrix**, at 240 light years. Next is blood-red **Betelgeuse** at the top left of Orion, 600 light years away (see December's Topic).

The constellation's brightest star, blue-white **Rigel**, is a vigorous young star more than twice as hot as our Sun, and 50,000 times as bright. Rigel lies 800 light years from us, roughly the same distance as the star that marks the other corner of Orion's tunic – **Saiph** – and the two outer stars of the belt, **Alnitak** (left) and **Mintaka** (right).

We travel 1300 light years from home to reach the middle star of the belt, **Alnilam**. And at the same distance, we reach the stars of the 'sword' hanging below the belt – the lair of the great **Orion Nebula**.

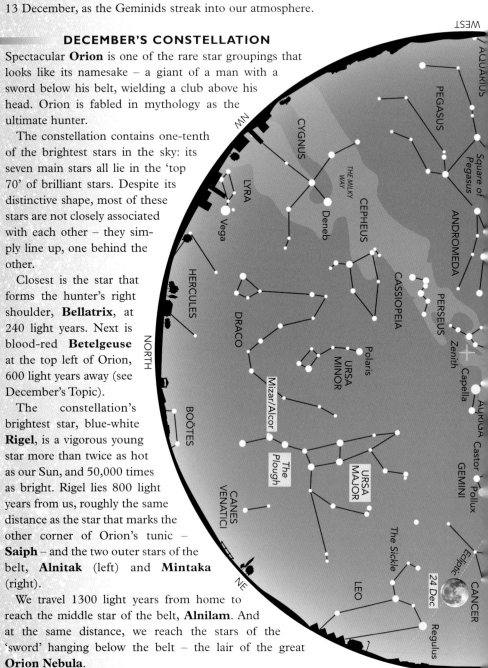

beginning of December and ? pm at the end of the month. The planets move slightly relative o the stars during the month.

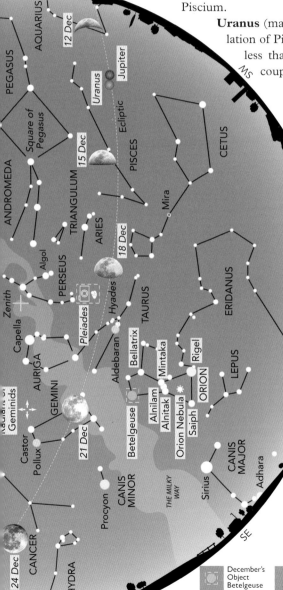

WEST

WEST

AQUARIUS

PEGASUS

12 Dec

Jupiter

Uranus

Ecliptic

Square of Pegasus

ANDROMEDA

15 Dec

PISCES

CETUS

Algol

TRIANGULUM

Mira

Zenith

PERSEUS

ARIES

18 Dec

Capella

Pleiades

Hyades

TAURUS

ERIDANUS

AURIGA

Aldebaran

Bellatrix

Mintaka

Rigel

LEPUS

Castor

GEMINI

Alnilam

Alnitak

ORION

Pollux

21 Dec

Betelgeuse

Orion Nebula

Saiph

COLUMBA

Radiants of Geminids

Procyon

CANIS MINOR

THE MILKY WAY

Sirius

CANIS MAJOR

Adhara

24 Dec

CANCER

HYDRA

SE

EAST

PLANETS ON VIEW

While the bright winter stars may be rising in the southeast, they are still no match for magnitude –2.4 **Jupiter**, over in the southwest and setting at around midnight. During the month, the giant planet moves gradually eastwards from Aquarius into Pisces. Binoculars or a small telescope will show its four largest moons – confusingly looking like five from 28 to 31 December as Jupiter moves in front of a star of similar brightness, 20 Piscium.

Uranus (magnitude 5.9) also lies in in the constellation of Pisces; you can easily spot it as the 'star' less than a degree above Jupiter on the last couple of nights of the year.

With a telescope, you'll find distant **Neptune** in Capricornus, at magnitude 7.9. It sets at around 9 pm.

Saturn, currently in Virgo, rises about 2 am and shines at magnitude 0.8 – just a little brighter than the constellation's main star, Spica, which lies directly below Saturn. A small telescope will reveal Saturn's amazing rings, and its biggest moon, Titan (magnitude 8.6) – but on the mornings of 10 and 11 December the planet appears to have garnered an even brighter moon as it passes the background star k Virginis (magnitude 5.8).

The Morning Star, **Venus**, rises around 4 am and shines at magnitude –4.6; one of these dark winter mornings would be a good time to check the theory that Venus is bright enough to cast shadows. A small

	December's Object Betelgeuse
	December's Picture Pleiades
	Radiant of Geminids

| Jupiter |
| Uranus |
| Moon |

MOON		
Date	**Time**	**Phase**
5	5.36 pm	New Moon
13	1.59 pm	First Quarter
21	8.13 am	Full Moon
28	4.18 am	Last Quarter

telescope shows its phase change from a crescent to half lit during the month.

Mercury is at greatest elongation on 1 December, but is too close to the Sun to be visible, as is **Mars**.

MOON

In the early morning of 1 December, the crescent Moon lies to the right of Saturn. On 2 December, it's near Venus in the dawn sky, with Spica above, while the following morning you'll find a slender crescent Moon below the Morning Star. The Moon is near Jupiter on 13 December. On the night of 18/19 December, the Moon passes just under the Pleiades. The morning of 29 December sees the Last Quarter Moon forming a triangle with Saturn (upper left) and Spica (lower left). The waning crescent Moon adorns the last morning sky of the year, 31 December, with the brilliant Morning Star Venus.

SPECIAL EVENTS

The maximum of the **Geminid** meteor shower falls on **13 December**. These shooting stars are debris shed from an asteroid called Phaethon and are therefore quite substantial, and thus bright. This year you'll be best observing them after midnight, when the Moon has set.

There's a total eclipse of the Moon in the early morning of **21 December**. But it won't be too spectacular as seen from the UK: the Moon moves into the Earth's shadow – low in the northwest – just as the sky is brightening. The best views will be from the Americas and the Pacific. The eclipse starts at 6.32 am and ends at 10.02 am, with totality between 7.40 and 8.53 am.

Later on the same day, **21 December**, the Winter Solstice occurs at 11.38 pm. As a result of the tilt of the Earth's axis, the Sun reaches its lowest point in the heavens as seen from the northern hemisphere: we get the shortest days, and the longest nights.

DECEMBER'S OBJECT

Known to generations of schoolkids as 'Beetle-Juice', **Betelgeuse** is one of the biggest stars known. If placed in the Solar System, it would swamp the planets all the way out to the asteroid belt. And it's one of just a few stars to be imaged as a visible disc from Earth.

Almost 1000 times wider than the Sun, Betelgeuse is a serious red giant – a star close to the end of its life. Its middle-age spread has been created by the dying nuclear reactions in its core, causing its outer layers swell and cool. The star also fluctuates slightly in brightness as it tries to get a grip on its billowing gases.

The star's vivid red colour has led to it attracting several names, including 'The Armpit of the Sacred One'!

⊙ *Viewing tip*
This is the month when you may be thinking of buying a telescope as a Christmas present for a budding stargazer. Beware! Unscrupulous mail-order catalogues selling 'gadgets' often advertise small telescopes that boast huge magnifications. This is known as 'empty magnification' – blowing up an image that the lens or mirror simply doesn't have the ability to get to grips with, so all you see is a bigger blur. A rule of thumb is to use a maximum magnification no greater than twice the diameter of your lens or mirror in millimetres. So if you have a 100 mm reflecting telescope, go no higher than 200X.

Betelgeuse will exit the Universe in a spectacular supernova explosion. As a result of the breakdown of nuclear reactions at its heart, the star will explode – and it will shine as brightly in our skies as the Moon.

DECEMBER'S PICTURE

The small-but-perfectly-formed **Pleiades** (M45) star cluster is as much a feature of our winter skies as the magnificent constellation of Orion. The 'Seven Sisters' are just a fraction of the 1000 stars making up the cluster, which lies around 440 light years away. Keen-sighted observers in dark conditions can spot over 12 Pleiads. And they are a delightful sight in binoculars.

As this image reveals, the stars are hot and blue – fledglings on the celestial age scale. They may have been born within the last 100 million years (compared to our Sun's age of 4.6 billion years).

The glorious nebulosity around the stars was once thought to be natal gas. But now it's reckoned that the Pleiades should have blown away their original nebula after 100 million years. The cluster is probably passing through a thick region of interstellar dust that reflects the stars' brilliant blue light.

This beautiful nest of stars will probably cling together for another 250 million years – after which the youngsters will go their own ways.

▲ *In this glorious image, taken in California, Michael Stecker captured the Pleiades and their delicate nebulosity perfectly. He took the photo through a 155mm f/7 AstroPhysics refractor on medium format ISO 400 colour film using a Pentax 6 x 7 camera. For maximum effect, he combined two separate 45-minute exposures.*

DECEMBER'S TOPIC
Double stars

The Sun is an exception in having singleton status. Over half the stars you see in the sky are paired up – they're double stars, or binaries. Look no further than the **Plough** in **Ursa Major** for a beautiful naked-eye example: the penultimate star in the Plough's handle (at the bend) is clearly double.

Mizar (the brighter star) and its fainter companion **Alcor** have often been named 'the horse and rider'. But until recently, there was dispute as to whether they were really in orbit around one another, or just stars that happened to lie in the same direction. Now, the sensitive Hipparcos satellite has pinned down their distances to 78.1 light years (Mizar) and 81.1 light years (Alcor). Although this is a gap of 3 light years, errors in the measurements could mean that the separation is only 0.7 light years, so the two stars could still form a binary system

There's always something to see in our Solar System, from planets to meteors or the Moon. These objects are very close to us – in astronomical terms – so their positions, shapes and sizes appear to change constantly. It is important to know when, where and how to look if you are to enjoy exploring Earth's neighbourhood. Here we give the best dates in 2010 for observing the planets and meteors (weather permitting!), and explain some of the concepts that will help you to get the most out of your observing.

THE INFERIOR PLANETS

A planet with an orbit that lies closer to the Sun than the orbit of Earth is known as *inferior*. Mercury and Venus are the inferior planets. They show a full range of phases (like the Moon) from the thinnest crescents to full, depending on their position in relation to the Earth and the Sun. The diagram below shows the various positions of the inferior planets. They are invisible when at *conjunction*, when they are behind the Sun or between the Earth and the Sun and lost in the latter's glare.

Magnitudes

Astronomers measure the brightness of stars, planets and other celestial objects using a scale of *magnitudes*. Somewhat confusingly, fainter objects have higher magnitudes, while brighter objects have lower magnitudes; the most brilliant stars have negative magnitudes! Naked-eye stars range from magnitude −1.5 for the brightest star, Sirius, to +6.5 for the faintest stars you can see on a really dark night. As a guide, here are the magnitudes of selected objects:

Sun	−26.7
Full Moon	−12.5
Venus (at its brightest)	−4.7
Sirius	−1.5
Betelgeuse	+0.4
Polaris (Pole Star)	+2.0
Faintest star visible to the naked eye	+6.5
Faintest star visible to the Hubble Space Telescope	+31

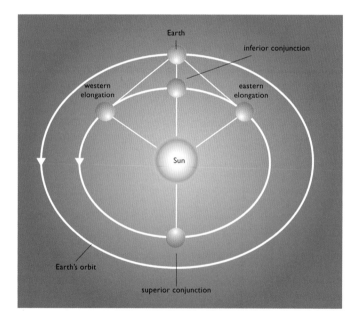

◀ At eastern or western elongation, an inferior planet is at its maximum angular distance from the Sun. Conjunction occurs at two stages in the planet's orbit. Under certain circumstances, an inferior planet can transit across the Sun's disc at inferior conjunction.

Mercury

In the second half of January Mercury is visible low in the southeast before sunrise; it reaches its greatest western elongation on 27 January. It re-emerges in the evening in late March for its best evening appearance of the year, but disappears again in mid-April. Its best morning appearance is in mid-September, but it's rapidly lost in the Sun's glare.

Maximum elongations of Mercury in 2010	
Date	Separation
27 January	24.8° west
8 April	19.4° east
26 May	25.1° west
7 August	27.4° east
19 September	17.9° west
1 December	21.5° east

Maximum elongation of Venus in 2010	
Date	Separation
20 August	46.0° east

Venus

Venus reaches its greatest eastern elongation of 2010 on 20 August, and towards the end of the year is approaching greatest western elongation. From March to August it is brilliant in the west as an Evening Star, disappearing from view by mid-September. It re-emerges as a Morning Star in November.

THE SUPERIOR PLANETS

The superior planets are those with orbits that lie beyond that of the Earth. They are Mars, Jupiter, Saturn, Uranus and Neptune. The best time to observe a superior planet is when the Earth lies between it and the Sun. At this point in a planet's orbit, it is said to be at *opposition*.

▶ *Superior planets are invisible at conjunction. At quadrature the planet is at right angles to the Sun as viewed from Earth. Opposition is the best time to observe a superior planet.*

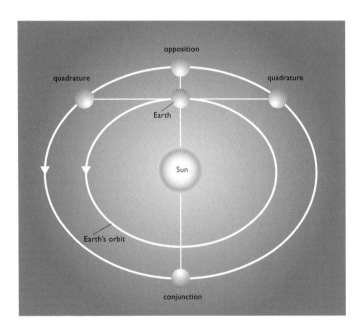

Progress of Mars through the constellations	
Early Jan	Leo
Mid-Jan – mid-May	Cancer
Mid-May – mid-July	Leo
Mid-July – late Sept	Virgo

Mars

Mars is at its best during the first half of the year. It is at opposition on 29 January, when it is visible for most of the night. It gradually drops back, until by September it is difficult to spot. It is lost from view for the rest of the year.

Jupiter

Jupiter is visible in the west during January, but is best placed for observing from late summer to the end of the year. During September it is the brightest object in Pisces, shining low in the south around midnight and visible all night long. It reaches opposition on 21 September.

Saturn

Saturn is in Virgo all year. It is at opposition on 22 March, and is at its best for observing from January to June. It is lost to view by September, but re-emerges in the morning towards the end of October.

Uranus

Uranus is best viewed from July onwards, when it lies on the border of Aquarius and Pisces, which it finally reaches near the end of the year. Visibility improves into the autumn, and it reaches opposition on 22 September, just a few hours after Jupiter, to which it sticks closely in the sky for much of the year.

Neptune

Neptune spends 2010 in Capricornus and is best viewed from June onwards. It reaches opposition on 20 August.

SOLAR AND LUNAR ECLIPSES

Solar Eclipses

There are two solar eclipses in 2010, on 15 January and 11 July. The former is annular and totality can be seen from parts of central Africa east across the Indian Ocean to India, Sri Lanka, Myanmar (Burma) and China, while the partially eclipsed Sun can be seen from most of Africa, Asia and southeastern Europe. The latter is total and visible from the South Pacific from the Cook Islands to the tip of South America. A partial eclipse will be seen in Chile and Argentina.

Lunar Eclipses

The partial lunar eclipse on 26 June will not be visible from the UK, while the total lunar eclipse of 21 December will occur around dawn as the Moon sets in the northwest.

> **Astronomical distances**
> For objects in the Solar System, such as the planets, we can give their distances from the Earth in kilometres. But the distances are just too huge once we reach out to the stars. Even the nearest star (Proxima Centauri) lies 25 million million kilometres away. So astronomers use a larger unit – the *light year*. This is the distance that light travels in one year, and it equals 9.46 million million kilometres. Here are the distances to some familiar astronomical objects, in light years:
>
> | | |
> |---|---|
> | Proxima Centauri | 4.2 |
> | Betelgeuse | 600 |
> | Centre of the Milky Way | 26,000 |
> | Andromeda Galaxy | 2.5 million |
> | Most distant galaxies seen by the Hubble Space Telescope | 13 billion |

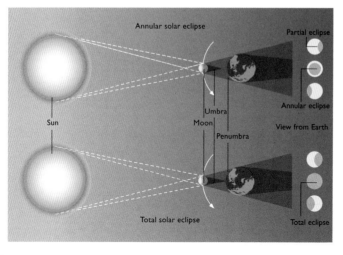

◀ *Where the dark central part (the umbra) of the Moon's shadow reaches the Earth, a total eclipse is seen. People located within the penumbra see a partial eclipse. If the umbral shadow does not reach Earth, an annular eclipse is seen. This type of eclipse occurs when the Moon is at a distant point in its orbit and is not quite large enough to cover the whole of the Sun's disc.*

Dates of maximum for selected meteor showers

Meteor shower	Date of maximum
Quadrantids	3/4 January
Lyrids	21/22 April
Eta Aquarids	4/5 May
Perseids	12/13 August
Orionids	20/21 October
Leonids	17/18 November
Geminids	13/14 December

▶ *Meteors from a common source, occurring during a shower, enter the atmosphere along parallel trajectories. As a result of perspective, however, they appear to diverge from a single point in the sky – the radiant.*

Angular separations

Astronomers measure the distance between objects, as we see them in the sky, by the angle between the objects in degrees (symbol °). From the horizon to the point above your head is 90 degrees. All around the horizon is 360 degrees.
You can use your hand, held at arm's length, as a rough guide to angular distances, as follows:
Width of index finger 1°
Width of clenched hand 10°
Thumb to little finger
  on outspread hand 20°
For smaller distances, astronomers divide the degree into 60 arcminutes (symbol '), and the arcminute into 60 arcseconds (symbol ").

METEOR SHOWERS

Shooting stars – or *meteors* – are tiny particles of interplanetary dust, known as *meteoroids*, burning up in the Earth's atmosphere. At certain times of year, the Earth passes through a stream of these meteoroids (usually debris left behind by a comet) and a *meteor shower* is seen. The point in the sky from which the meteors appear to emanate is known as the *radiant*. Most showers are known by the constellation in which the radiant is situated.

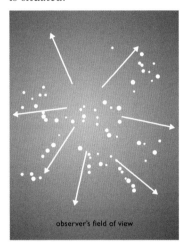

observer's field of view

When watching meteors for a coordinated meteor programme, observers generally note the time, seeing conditions, cloud cover, their own location, the time and brightness of each meteor, and whether it was from the main meteor stream. It is also worth noting details of persistent afterglows (trains) and fireballs, and making counts of how many meteors appear in a given period.

COMETS

Comets are small bodies in orbit about the Sun. Consisting of frozen gases and dust, they are often known as 'dirty snowballs'. When their orbits bring them close to the Sun, the ices evaporate and dramatic tails of gas and dust can sometimes be seen.

A number of comets move round the Sun in fairly small, elliptical orbits in periods of a few years; others have much longer periods. Most really brilliant comets have orbital periods of several thousands or even millions of years. The exception is Comet Halley, a bright comet with a period of about 76 years. It was last seen with the naked eye in 1986.

Binoculars and wide-field telescopes provide the best views of comet tails. Larger telescopes with a high magnification are necessary to observe fine detail in the gaseous head (*coma*). Most comets are discovered with professional instruments, but a few are still found by experienced amateur astronomers.

Comet Hartley 2 is the only known comet predicted to reach naked-eye brightness in 2010, during October and early November. However, there's always a chance of a bright new comet putting in a surprise appearance.

Deep-sky objects are 'fuzzy patches' that lie outside the Solar System. They include star clusters, nebulae and galaxies. To observe the majority of deep-sky objects you will need binoculars or a telescope, but there are also some beautiful naked-eye objects, notably the Pleiades and the Orion Nebula.

The faintest object that an instrument can see is its *limiting magnitude*. The table gives a rough guide, for good seeing conditions, for a variety of small- to medium-sized telescopes.

We have provided a selection of recommended deep-sky targets, together with their magnitudes. Some are described in more detail in our 'Object of the Month' features. Look on the appropriate month's map to find which constellations are on view, and then choose your objects using the list below. We have provided celestial coordinates for readers with detailed star maps. The suggested times of year for viewing are when the constellation is highest in the sky in the late evening.

Limiting magnitude for small to medium telescopes	
Aperture (mm)	Limiting magnitude
50	+11.2
60	+11.6
70	+11.9
80	+12.2
100	+12.7
125	+13.2
150	+13.6

RECOMMENDED DEEP-SKY OBJECTS

Andromeda – autumn and early winter

M31 (NGC 224) Andromeda Galaxy	3rd-magnitude spiral galaxy RA 00h 42.7m Dec +41° 16'
M32 (NGC 221)	8th-magnitude elliptical galaxy, a companion to M31 RA 00h 42.7m Dec +40° 52'
M110 (NGC 205)	8th-magnitude elliptical galaxy RA 00h 40.4m Dec +41° 41'
NGC 7662 Blue Snowball	8th-magnitude planetary nebula RA 23h 25.9m Dec +42° 33'

Aquarius – late autumn and early winter

M2 (NGC 7089)	6th-magnitude globular cluster RA 21h 33.5m Dec –00° 49'
M72 (NGC 6981)	9th-magnitude globular cluster RA 20h 53.5m Dec –12° 32'
NGC 7293 Helix Nebula	7th-magnitude planetary nebula RA 22h 29.6m Dec –20° 48'
NGC 7009 Saturn Nebula	8th-magnitude planetary nebula RA 21h 04.2m Dec –11° 22'

Aries – early winter

NGC 772	10th-magnitude spiral galaxy RA 01h 59.3m Dec +19° 01'

Auriga – winter

M36 (NGC 1960)	6th-magnitude open cluster RA 05h 36.1m Dec +34° 08'
M37 (NGC 2099)	6th-magnitude open cluster RA 05h 52.4m Dec +32° 33'
M38 (NGC 1912)	6th-magnitude open cluster RA 05h 28.7m Dec +35° 50'

Cancer – late winter to early spring

M44 (NGC 2632) Praesepe or Beehive	3rd-magnitude open cluster RA 08h 40.1m Dec +19° 59'
M67 (NGC 2682)	7th-magnitude open cluster RA 08h 50.4m Dec +11° 49'

Canes Venatici – visible all year

M3 (NGC 5272)	6th-magnitude globular cluster RA 13h 42.2m Dec +28° 23'

M51 (NGC 5194/5) Whirlpool Galaxy	8th-magnitude spiral galaxy RA 13h 29.9m Dec +47° 12'
M63 (NGC 5055)	9th-magnitude spiral galaxy RA 13h 15.8m Dec +42° 02'
M94 (NGC 4736)	8th-magnitude spiral galaxy RA 12h 50.9m Dec +41° 07'
M106 (NGC4258)	8th-magnitude spiral galaxy RA 12h 19.0m Dec +47° 18'

Canis Major – late winter

M41 (NGC 2287)	4th-magnitude open cluster RA 06h 47.0m Dec –20° 44'

Capricornus – late summer and early autumn

M30 (NGC 7099)	7th-magnitude globular cluster RA 21h 40.4m Dec –23° 11'

Cassiopeia – visible all year

M52 (NGC 7654)	6th-magnitude open cluster RA 23h 24.2m Dec +61° 35'
M103 (NGC 581)	7th-magnitude open cluster RA 01h 33.2m Dec +60° 42'
NGC 225	7th-magnitude open cluster RA 00h 43.4m Dec +61 47'
NGC 457	6th-magnitude open cluster RA 01h 19.1m Dec +58° 20'
NGC 663	Good binocular open cluster RA 01h 46.0m Dec +61° 15'

Cepheus – visible all year

Delta Cephei	Variable star, varying between +3.5 and +4.4 with a period of 5.37 days. It has a magnitude +6.3 companion and they make an attractive pair for small telescopes or binoculars.

Cetus – late autumn

Mira (omicron Ceti)	Irregular variable star with a period of roughly 330 days and a range between +2.0 and +10.1.
M77 (NGC 1068)	9th-magnitude spiral galaxy RA 02h 42.7m Dec –00° 01'

Coma Berenices – spring

M53 (NGC 5024)	8th-magnitude globular cluster *RA 13h 12.9m Dec +18° 10'*
M64 (NGC 4286) Black Eye Galaxy	8th-magnitude spiral galaxy with a prominent dust lane that is visible in larger telescopes. *RA 12h 56.7m Dec +21° 41'*
M85 (NGC 4382)	9th-magnitude elliptical galaxy *RA 12h 25.4m Dec +18° 11'*
M88 (NGC 4501)	10th-magnitude spiral galaxy *RA 12h 32.0m Dec.+14° 25'*
M91 (NGC 4548)	10th-magnitude spiral galaxy *RA 12h 35.4m Dec +14° 30'*
M98 (NGC 4192)	10th-magnitude spiral galaxy *RA 12h 13.8m Dec +14° 54'*
M99 (NGC 4254)	10th-magnitude spiral galaxy *RA 12h 18.8m Dec +14° 25'*
M100 (NGC 4321)	9th-magnitude spiral galaxy *RA 12h 22.9m Dec +15° 49'*
NGC 4565	10th-magnitude spiral galaxy *RA 12h 36.3m Dec +25° 59'*

Cygnus – late summer and autumn

Cygnus Rift	Dark cloud just south of Deneb that appears to split the Milky Way in two.
NGC 7000 North America Nebula	A bright nebula against the background of the Milky Way, visible with binoculars under dark skies. *RA 20h 58.8m Dec +44° 20'*
NGC 6992 Veil Nebula (part)	Supernova remnant, visible with binoculars under dark skies. *RA 20h 56.8m Dec +31 28'*
M29 (NGC 6913)	7th-magnitude open cluster *RA 20h 23.9m Dec +36° 32'*
M39 (NGC 7092)	Large 5th-magnitude open cluster *RA 21h 32.2m Dec +48° 26'*
NGC 6826 Blinking Planetary	9th-magnitude planetary nebula *RA 19 44.8m Dec +50° 31'*

Delphinus – late summer

NGC 6934	9th-magnitude globular cluster *RA 20h 34.2m Dec +07° 24'*

Draco – midsummer

NGC 6543	9th-magnitude planetary nebula *RA 17h 58.6m Dec +66° 38'*

Gemini – winter

M35 (NGC 2168)	5th-magnitude open cluster *RA 06h 08.9m Dec +24° 20'*
NGC 2392 Eskimo Nebula	8–10th-magnitude planetary nebula *RA 07h 29.2m Dec +20° 55'*

Hercules – early summer

M13 (NGC 6205)	6th-magnitude globular cluster *RA 16h 41.7m Dec +36° 28'*
M92 (NGC 6341)	6th-magnitude globular cluster *RA 17h 17.1m Dec +43° 08'*
NGC 6210	9th-magnitude planetary nebula *RA 16h 44.5m Dec +23 49'*

Hydra – early spring

M48 (NGC 2548)	6th-magnitude open cluster *RA 08h 13.8m Dec –05° 48'*
M68 (NGC 4590)	8th-magnitude globular cluster *RA 12h 39.5m Dec –26° 45'*

M83 (NGC 5236)	8th-magnitude spiral galaxy *RA 13h 37.0m Dec –29° 52'*
NGC 3242 Ghost of Jupiter	9th-magnitude planetary nebula *RA 10h 24.8m Dec –18°38'*

Leo – spring

M65 (NGC 3623)	9th-magnitude spiral galaxy *RA 11h 18.9m Dec +13° 05'*
M66 (NGC 3627)	9th-magnitude spiral galaxy *RA 11h 20.2m Dec +12° 59'*
M95 (NGC 3351)	10th-magnitude spiral galaxy *RA 10h 44.0m Dec +11° 42'*
M96 (NGC 3368)	9th-magnitude spiral galaxy *RA 10h 46.8m Dec +11° 49'*
M105 (NGC 3379)	9th-magnitude elliptical galaxy *RA 10h 47.8m Dec +12° 35'*

Lepus – winter

M79 (NGC 1904)	8th-magnitude globular cluster *RA 05h 24.5m Dec –24° 33'*

Lyra – spring

M56 (NGC 6779)	8th-magnitude globular cluster *RA 19h 16.6m Dec +30° 11'*
M57 (NGC 6720) Ring Nebula	9th-magnitude planetary nebula *RA 18h 53.6m Dec +33° 02'*

Monoceros – winter

M50 (NGC 2323)	6th-magnitude open cluster *RA 07h 03.2m Dec –08° 20'*
NGC 2244	Open cluster surrounded by the faint Rosette Nebula, NGC 2237. Visible in binoculars. *RA 06h 32.4m Dec +04° 52'*

Ophiuchus – summer

M9 (NGC 6333)	8th-magnitude globular cluster *RA 17h 19.2m Dec –18° 31'*
M10 (NGC 6254)	7th-magnitude globular cluster *RA 16h 57.1m Dec –04° 06'*
M12 (NCG 6218)	7th-magnitude globular cluster *RA 16h 47.2m Dec –01° 57'*
M14 (NGC 6402)	8th-magnitude globular cluster *RA 17h 37.6m Dec –03° 15'*
M19 (NGC 6273)	7th-magnitude globular cluster *RA 17h 02.6m Dec –26° 16'*
M62 (NGC 6266)	7th-magnitude globular cluster *RA 17h 01.2m Dec –30° 07'*
M107 (NGC 6171)	8th-magnitude globular cluster *RA 16h 32.5m Dec –13° 03'*

Orion – winter

M42 (NGC 1976) Orion Nebula	4th-magnitude nebula *RA 05h 35.4m Dec –05° 27'*
M43 (NGC 1982)	5th-magnitude nebula *RA 05h 35.6m Dec –05° 16'*
M78 (NGC 2068)	8th-magnitude nebula *RA 05h 46.7m Dec +00° 03'*

Pegasus – autumn

M15 (NGC 7078)	6th-magnitude globular cluster *RA 21h 30.0m Dec +12° 10'*

Perseus – autumn to winter

M34 (NGC 1039)	5th-magnitude open cluster *RA 02h 42.0m Dec +42° 47'*
M76 (NGC 650/1) Little Dumbbell	11th-magnitude planetary nebula *RA 01h 42.4m Dec +51° 34'*

NGC 869/884 Double Cluster	Pair of open star clusters RA 02h 19.0m Dec +57° 09' RA 02h 22.4m Dec +57° 07'

Pisces – autumn

M74 (NGC 628)	9th-magnitude spiral galaxy RA 01h 36.7m Dec +15° 47'

Puppis – late winter

M46 (NGC 2437)	6th-magnitude open cluster RA 07h 41.8m Dec −14° 49'
M47 (NGC 2422)	4th-magnitude open cluster RA 07h 36.6m Dec −14° 30'
M93 (NGC 2447)	6th-magnitude open cluster RA 07h 44.6m Dec −23° 52'

Sagitta – late summer

M71 (NGC 6838)	8th-magnitude globular cluster RA 19h 53.8m Dec +18° 47'

Sagittarius – summer

M8 (NGC 6523) Lagoon Nebula	6th-magnitude nebula RA 18h 03.8m Dec −24° 23'
M17 (NGC 6618) Omega Nebula	6th-magnitude nebula RA 18h 20.8m Dec −16° 11'
M18 (NGC 6613)	7th-magnitude open cluster RA 18h 19.9m Dec −17 08'
M20 (NGC 6514) Trifid Nebula	9th-magnitude nebula RA 18h 02.3m Dec −23° 02'
M21 (NGC 6531)	6th-magnitude open cluster RA 18h 04.6m Dec −22° 30'
M22 (NGC 6656)	5th-magnitude globular cluster RA 18h 36.4m Dec −23° 54'
M23 (NGC 6494)	5th-magnitude open cluster RA 17h 56.8m Dec −19° 01'
M24 (NGC 6603)	5th-magnitude open cluster RA 18h 16.9m Dec −18° 29'
M25 (IC 4725)	5th-magnitude open cluster RA 18h 31.6m Dec −19° 15'
M28 (NGC 6626)	7th-magnitude globular cluster RA 18h 24.5m Dec −24° 52'
M54 (NGC 6715)	8th-magnitude globular cluster RA 18h 55.1m Dec −30° 29'
M55 (NGC 6809)	7th-magnitude globular cluster RA 19h 40.0m Dec −30° 58'
M69 (NGC 6637)	8th-magnitude globular cluster RA 18h 31.4m Dec −32° 21'
M70 (NGC 6681)	8th-magnitude globular cluster RA 18h 43.2m Dec −32° 18'
M75 (NGC 6864)	9th-magnitude globular cluster RA 20h 06.1m Dec −21° 55'

Scorpius (northern part) – midsummer

M4 (NGC 6121)	6th-magnitude globular cluster RA 16h 23.6m Dec −26° 32'
M7 (NGC 6475)	3rd-magnitude open cluster RA 17h 53.9m Dec −34° 49'
M80 (NGC 6093)	7th-magnitude globular cluster RA 16h 17.0m Dec −22° 59'

Scutum – mid to late summer

M11 (NGC 6705) Wild Duck Cluster	6th-magnitude open cluster RA 18h 51.1m Dec −06° 16'

M26 (NGC 6694)	8th-magnitude open cluster RA 18h 45.2m Dec −09° 24'

Serpens – summer

M5 (NGC 5904)	6th-magnitude globular cluster RA 15h 18.6m Dec +02° 05'
M16 (NGC 6611)	6th-magnitude open cluster, surrounded by the Eagle Nebula RA 18h 18.8m Dec −13° 47'

Taurus – winter

M1 (NGC 1952) Crab Nebula	8th-magnitude supernova remnant RA 05h 34.5m Dec +22° 00'
M45 Pleiades	1st-magnitude open cluster, an excellent binocular object. RA 03h 47.0m Dec +24° 07'

Triangulum – autumn

M33 (NGC 598)	6th-magnitude spiral galaxy RA 01h 33.9m Dec +30° 39'

Ursa Major – all year

M81 (NGC 3031)	7th-magnitude spiral galaxy RA 09h 55.6m Dec +69° 04'
M82 (NGC 3034)	8th-magnitude starburst galaxy RA 09h 55.8m Dec +69° 41'
M97 (NGC 3587) Owl Nebula	12th-magnitude planetary nebula RA 11h 14.8m Dec +55° 01'
M101 (NGC 5457)	8th-magnitude spiral galaxy RA 14h 03.2m Dec +54° 21'
M108 (NGC 3556)	10th-magnitude spiral galaxy RA 11h 11.5m Dec +55° 40'
M109 (NGC 3992)	10th-magnitude spiral galaxy RA 11h 57.6m Dec +53° 23'

Virgo – spring

M49 (NGC 4472)	8th-magnitude elliptical galaxy RA 12h 29.8m Dec +08° 00'
M58 (NGC 4579)	10th-magnitude spiral galaxy RA 12h 37.7m Dec +11° 49'
M59 (NGC 4621)	10th-magnitude elliptical galaxy RA 12h 42.0m Dec +11° 39'
M60 (NGC 4649)	9th-magnitude elliptical galaxy RA 12h 43.7m Dec +11° 33'
M61 (NGC 4303)	10th-magnitude spiral galaxy RA 12h 21.9m Dec +04° 28'
M84 (NGC 4374)	9th-magnitude elliptical galaxy RA 12h 25.1m Dec +12° 53'
M86 (NGC 4406)	9th-magnitude elliptical galaxy RA 12h 26.2m Dec +12° 57'
M87 (NGC 4486)	9th-magnitude elliptical galaxy RA 12h 30.8m Dec +12° 24'
M89 (NGC 4552)	10th-magnitude elliptical galaxy RA 12h 35.7m Dec +12° 33'
M90 (NGC 4569)	9th-magnitude spiral galaxy RA 12h 36.8m Dec +13° 10'
M104 (NGC 4594) Sombrero Galaxy	Almost edge-on 8th-magnitude spiral galaxy. RA 12h 40.0m Dec −11° 37'

Vulpecula – late summer and autumn

M27 (NGC 6853) Dumbbell Nebula	8th-magnitude planetary nebula RA 19h 59.6m Dec +22° 43'

▲ *Reflector and refractor compared. The 130 mm reflector on the left has its mirror at the bottom of the tube, and a secondary mirror assembly at the top of the tube to reflect the image to the eyepiece at upper left. The 80 mm refractor at right has a lens at the top of the tube with the eyepiece at the bottom.*

▼ *Celestron NexStar SLT 102 mm refractor (left) compared with NexStar SLT 130 mm reflector.*

Reflector or refractor?

Probably the first thing than any newcomer to astronomy learns about telescopes is that there are two basic types – refractors and reflectors. Refracting telescopes use a lens at the top of the tube to focus the light, while reflecting telescopes use a dish-shaped mirror at the bottom instead. So if you want to buy your first telescope, which should you consider?

Leaving aside all the technical considerations, cost alone isn't a major factor. Take the Celestron range, for example. Their popular NexStar SLT range includes both types, at broadly similar prices between £200 and £300 for telescopes with GO TO computer control. Alternatively, Sky-Watcher have a basic 130 mm reflector in the same price range as a 90 mm refractor. In each case, you get a larger reflector for your money than a refractor.

So that might seem to be the end to it. Why buy a smaller telescope when you can get a larger one for less money? But as you might imagine, there's more to it than simply the aperture of the instrument.

Pros and cons

The very first refracting telescopes left a lot to be desired. They suffered badly from false colour, as the lens behaved more like a prism and split up the incoming light into the colours of the rainbow. It was not until 1758, when John Dollond patented an achromatic lens, that telescopes really became usable. Achromatic implies freedom from colour, and all modern refractors other than toys are a variant of Dollond's design, which uses a combination of two separate glass elements of different types to correct for false colour. But the basic achromatic lens can only do a limited job of correction, and the widely available entry-level telescopes still have a trace of false colour. Basically, they bring green light to a single focal point, but light at the far ends of the spectrum focuses at a slightly different point.

In their favour, refractors of this sort are very reliable, need little maintenance and can give crisp, high-contrast images straight out of the box, time after time. Reflectors, by comparison, are more finicky.

A mirror has no false colour at all, but there are drawbacks with reflectors. One is that the mirror's reflective surface is very delicate and won't stand a lot of cleaning. The other main drawback is that because the mirror focuses its light at the top of the tube, you need an additional flat mirror within the tube itself to reflect the light to the side of the tube where you can view it. This second mirror causes some light loss, but its presence in the light path also affects the quality of the image itself somewhat, resulting in a slight loss of contrast of fine detail.

When it comes to the crunch it's often other practical factors that play a part. The mirror coatings of reflectors can deteriorate after a while, and cleaning them is tricky. The optical system can also get out of line, requiring realignment (called collimation). This is not a big job when you are used to it, but it can frighten beginners. Another problem with reflectors is that they can take longer to cool down when taken out into the cold night air from a warm room. Even a refractor still takes time to cool down, however, and in general you should take any telescope outside well before you begin observing.

▲ *Children observing with telescopes. At left, an 80 mm refractor; at right, a 130 mm reflector; the girl at the back is using 10 × 30 binoculars.*

However, refractors don't have it all their own way. One constant problem, particularly in the UK, is dewing up of the optics. The lens of a refractor is generally protected to a certain extent by an additional tube that projects beyond the lens itself, known as a dew shield. The aim of this is to prevent the glass of the lens from seeing any more of the night sky than necessary. Not only does this help to shield it from any lighting that happens to be around, but more importantly it helps to keep the surface of the lens above the dew point.

The dew point is the temperature at which moisture will condense out of the air. Once this happens, a glass surface will mist up straight away, so going below the dew point is fatal. The night sky is very effective at cooling things down, because it is at a very low temperature, even on a warm night. The aim of the dew shield is to limit the lens's exposure to the night sky to only the area being observed. A mirror, however, is generally much deeper inside its tube and rarely gets cold enough to dew up, which is a point in favour of reflectors.

The larger the telescope, the longer the tube, and in the case of large refractors this can make the instrument quite hard to use. To avoid the observer having to lie on the ground to view through them, almost every refractor needs what's called a star diagonal – a unit that reflects the light through 90° so that you can view at a more comfortable angle. This inverts the image, so the Moon appears with its features back to front, for example, and star fields are a mirror image of the view on a star map. It also involves an extra reflection, which wastes a little light, so one of the advantages of a refractor is nulled.

▶ *Even using a star diagonal, observing objects high in the sky with a large refractor – here a 150 mm – can mean crouching low down.*

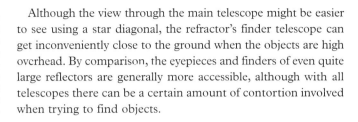

**SUPPLIERS OF
THE EQUIPMENT
SHOWN HERE**

Celestron from
David Hinds Ltd,
Leighton Buzzard, Beds:
www.celestron.uk.com

Sky-Watcher from
Optical Vision Ltd:
www.opticalvision.co.uk

Baader Fringe-Killer from
Baader Planetarium:
www.baader-planetarium.de/
– available from most
suppliers of astronomical
equipment.

Prices quoted here are
current as of mid-2009.

Although the view through the main telescope might be easier to see using a star diagonal, the refractor's finder telescope can get inconveniently close to the ground when the objects are high overhead. By comparison, the eyepieces and finders of even quite large reflectors are generally more accessible, although with all telescopes there can be a certain amount of contortion involved when trying to find objects.

The view through the eyepiece

So if you could view the same object through a reflector and a refractor, would you notice any differences? At first glance, probably not. Take your first look at, say, the Moon, using the standard eyepiece supplied with the instrument, which gives a fairly low magnification, and both views would be stunning. The Moon's craters and mountains should stand out crisply against the black background. A star cluster will appear similarly well defined. It's not until you look closely, at higher magnifications, that differences start to appear.

In the refractor, the edge of the Moon has a coloured fringe that becomes more obvious as you increase the magnification, although the details within this should be sharp. At higher magnifications individual craters may appear to have blue edges on one side, and yellowish ones on the other. In the reflector, however, no colour is obvious, particularly when the Moon is high in the sky and free from any colour introduced by layers in our own atmosphere. When any object is low in the sky, atmospheric refraction does create its own false colours, whatever telescope you are observing with.

A planet, such as Saturn or Jupiter, appears to have a slight coloured haze around it when seen through a basic refractor, even though the details of the planet are sharp. This haze is often bluish or purplish. One way to get rid of it is to use a straw-coloured filter in the eyepiece, such as the Baader Fringe-Killer. This can get rid of the worst of the purple haze and, although it reduces the image brightness somewhat and gives a yellowish tinge to objects, because there is less unwanted light smearing out the view it can help you to see finer details on the Moon and planets.

Apochromatic refractors

All the preceding remarks about refractors apply to the ordinary instruments comparable in price with similar-sized reflectors. But in recent years another breed of refractor, which has much better colour correction than run-of-the-mill instruments, has become popular. These are known either as apochromatic (APOs for short) fluorite or ED refractors, ED standing for Extra Dispersion. This refers to the material from which the lens elements are made. Apochromats have been around for a long time, but only with the introduction of specialized lens materials have they become at all affordable. Even so, an 80 mm ED refractor will set you back more for the tube alone, without mounting, than a fully mounted 200 mm reflector. Nevertheless, they are popular because of their convenience and optical performance

Such refractors also have a wider sharp field of view than the equivalent reflector. The very best, usually in the larger sizes, are known as astrographs and can give stunning performance.

Recent years have seen the resurgence of interest in very small refractors of high quality. The humble 60 mm refractor was long regarded as no more than a beginner's instrument, but these days apochromatic 66 mm refractors are being used for high-definition wide-field astrophotography and for bird- and nature-watching.

Your choice

If you want a small and convenient telescope for occasional viewing, a refractor of 80 mm to 120 mm aperture is fine. It will give you years of service and will show you a wide range of objects and some details on the bright planets. It will require no maintenance as long as you keep it out of the damp and dust, and will work on daytime objects as well, though you'll need to use a diagonal to avoid an upside-down image. Larger refractors require more space and tend to be more suited to the devoted observers.

Observers in towns might find that a refractor suits them better than a reflector, as the dew shield offers better protection against stray light. Many planetary observers prefer refractors, as do Sun observers. You should never observe the Sun directly, as it is bright enough to blind you instantly, but by using approved solar filters over the full aperture of the telescope, it's possible to get excellent views in safety. Where the Sun is concerned, apertures need be no larger than 100 mm as there's plenty of light available.

Reflectors tend to be for the more committed observer, who is prepared to put up with their foibles in return for images that are free from false colour and, generally speaking, brighter than those of a similar-priced refractor. Deep-sky observers, or those living in the country where the darker skies will permit fainter objects to be seen, will probably benefit from the greater light-gathering power of a larger reflector.

Whatever telescope you choose, the good news is that most telescopes from reputable manufacturers these days will perform well, and will provide you with your own window on the Universe.

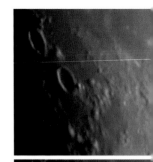

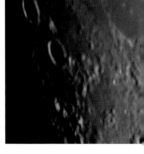

▲ *Comparison of reflector and refractor views of Mare Humorum.*

▼ *The Rosette Nebula in Monoceros, photographed with an 80 mm refractor from Edgware, in the suburbs of London. It was compiled from a total of of over three hours of separate exposures through narrowband colour filters using a monochrome CCD camera.*